Universitext

Universitext

Universitext is a series of textbooks that presents material from a wide variety of mathematical disciplines at master's level and beyond. The books, often well class-tested by their author, may have an informal, personal, even experimental approach to their subject matter. Some of the most successful and established books in the series have evolved through several editions, always following the evolution of teaching curricula, into very polished texts.

Thus as research topics trickle down into graduate-level teaching, first textbooks written for new, cutting-edge courses may make their way into *Universitext*.

More information about this series at http://www.springer.com/series/223

Jean-François Collet

Discrete Stochastic Processes and Applications

 Springer

Jean-François Collet
Laboratoire J.A. Dieudonné
Université de Nice Sophia-Antipolis
Nice Cedex 02
France

ISSN 0172-5939   ISSN 2191-6675 (electronic)
Universitext
ISBN 978-3-319-74017-1   ISBN 978-3-319-74018-8 (eBook)
https://doi.org/10.1007/978-3-319-74018-8

Library of Congress Control Number: 2017964594

Mathematics Subject Classification (2010): Primary: 60J10, 60J27, 60J28, 60J75, 94A17; Secondary: 26B25, 60J80, 60J85, 94A24

Printed on acid-free paper

This Springer imprint is published by the registered company Springer International Publishing AG part of Springer Nature
The registered company address is: Gewerbestrasse 11, 6330 Cham, Switzerland

Pour vous quatre

Preface

Stochastic processes are quite prevalent in scientific modeling, and the fine analysis of their mathematical properties relies on tools coming from such diverse branches of mathematics as (aside from probability theory of course) linear algebra, convex analysis, and information theory.

The aim of this book is to give a self-contained introduction to this extensive toolbox at a rather elementary level and to discuss its use in the study of Markov processes as well as in some other applications such as coding theory, population dynamics, and the design of search engines.

The first part of the book focuses on the rigorous theory of Markov chains and could provide a basis for a one-semester introductory course on this topic.

The main goal is to make it possible for the reader to develop a good intuition for the probabilistic concepts relevant to Markov processes without having to digest measure theory first. This is why we restrict ourselves to Markov processes on countable spaces, a.k.a. Markov chains. In this case, the mathematical form of the idea of absence of long-term memory is rather intuitive and does not require the extra technical tools (such as stopping times and filtrations) needed in the continuous case. The approach will be very gradual, starting with the discrete-time case, then moving on to continuous time. Along the way, new technicalities arise but the underlying probabilistic concepts are the same.

One of the most important results from the theory of Markov chains is the convergence theorem for ergodic chains (Theorem 1.39 in the text). This will be proved twice, first with probabilistic techniques in Chapter 1 (the coupling method is fundamentally probabilistic in nature) and second (for finite spaces) by pure linear algebra in Chapter 2. The fact that such a general result may be proved by attacking it from such radically different directions should cause readers to marvel once they have reached the end of Chapter 2, and it says a lot about the richness of the field and the beauty of mathematics. Chapter 2 ends with a popular application of the convergence result for Markov chains, the PageRank algorithm used in search engines.

The next step is to move on to continuous-time processes. The study of the Poisson process in Chapter 3 will serve as a pretext to introduce some of the relevant concepts without getting too technical. This will begin with a handmade construction of the Poisson process, before a review of the classical definitions and proof of their equivalence. Chapter 4 then deals with the rigorous theory of general continuous-time Markov processes. Since this text is intended as only a first exposition to stochastic processes, in order to avoid technicalities (which would mean introducing stopping times and filtrations) we do not cover the strong Markov property. Admittedly, the price to pay is that the rigorous proofs based on the use of skeletons may get a bit technical in places, so the reader who is interested mostly in applications might skip the detailed proofs.

Chapter 5 should provide some necessary relief after so much theory. It explores two kinds of examples: random walks and birth–death processes. The approach taken is much more applied than in the previous chapters and focuses on actual computations rather than theoretical results.

The first chapters provide two approaches to the study of the asymptotic behavior of Markov chains; a third route (and probably the least represented in textbooks on stochastic processes) is the use of entropy.

The second part of the text provides an introduction to various uses of convexity in probability as well as to the concept of entropy. It is more applied in nature (although mathematically rigorous) and might serve as a one-semester gentle introduction to coding and information theory.

At the heart of the uses of entropy is a certain number of inequalities that are in fact more or less all convexity inequalities. Assuming no prior knowledge of convexity whatsoever, Chapter 6 is an introduction to convex sets and convex maps, and to the relevant inequalities. In particular, it includes a discussion of Bregman divergences, which although they are now used in data clustering, are not often found in basic probability textbooks.

Chapter 7 collects the main quantities used in information theory; according to the expositional philosophy of the first part we present everything in the framework of discrete distributions, but the extension to absolutely continuous distributions is trivial.

The main theme here is that many quantities (such as Bregman divergences and Φ-divergences and in particular the Kullback–Leibler divergence) defined on the space of probability vectors may be used as distances even though they are not distances in the true mathematical sense of the term. What is meant here by "may be used as distances" becomes clear once the main inequalities, Jensen's and Pinsker's, are understood. In Section 7.4 we show how to use entropy to give a third proof of the convergence theorem for ergodic chains (at which point the above comment on the beauty and unity of mathematics should be reiterated). This is a use in the setting of Markov chains of what is known as the "entropy dissipation method" in partial differential equations and dynamical systems. The chapter ends with a detailed introduction to exponential families, which are parametric families of probability distributions that under some constraints maximize entropy. Finally, Chapter 8 explores the connection with binary coding; essentially, it is shown that the entropy of a random source provides an incompressible lower bound for the length of any uniquely decipherable code. To make a long story short, loss of memory in a Markov chain or complexity of a random source may be quantified using the same tool, entropy.

The prerequisites for this text are as follows.

- From linear algebra: multiplication of matrices, scalar product, and the concept of eigenvalue (no knowledge of diagonalization results is needed); the section on the Perron–Frobenius theorem is totally self-contained.
- From calculus: sequences and proofs by induction, manipulation of convergent series, the chain rule, computation of one-dimensional integrals by change of variables and integration by parts. From calculus in several variables, the notions of gradient and Hessian.
- From probability theory: discrete and absolutely continuous random variables; conditional probability, computations of expectations, moments, Gaussian variables.

Acknowledgments. Now comes the most pleasant and easiest paragraph of this book to write. It is indeed a pleasure to thank all the people without whom this book would not have been written: Didier Auroux, Florent Berthelin, Jaques Blum, Cédric Boulbe, David Chiron, Jean-Antoine Désidéri, Frédéric Gruy, Bernard Guy, David Hoff, Stéphane Junca, Frédéric Poupaud, Francesca Rapetti, Vitaly Volpert.

Touët-sur-Var
Jean-François Collet July 2017

Contents

Contents

Contents

Terminology, Abbreviations, and Typographical Conventions

Terminology and Abbreviations

Throughout the text, the words "positive" and "negative" are understood strictly; the components of a vector q are interchangeably denoted by q_i or $q(i)$, and similarly, the components of a matrix p are interchangeably denoted by $p_{i,j}$ or $p(i,j)$. We refer to the probability mass function of a discrete random variable X simply as its probability vector q, and then write $q(x)$ for $P(X = x)$, even in the case that X takes its values in an infinite countable set.

Unless otherwise stated, vectors appearing in products are understood as row vectors; therefore, in a product of a vector by a matrix, the vector will appear to the left of the matrix.

The end of a rigorous proof is indicated by the symbol $\square$. The notation := indicates that an equality should be understood as a definition of its left-hand-side (as opposed to equalities resulting from a computation).

Notation

- A^t: the transpose matrix of a matrix A, meaning $A^t(i,j) := A(j,i)$. This notation includes vectors, so for example, if x is a column vector, then x^t is the corresponding row vector.
- I_n: the $n \times n$ identity matrix; sometimes denoted by I when n is implicit.
- $\delta_{x,y}$ is the well-known Kronecker delta : it is 1 if $x = y$, and 0 otherwise.
- $\mathbb{N}^*$, $\mathbb{R}^*$: positive integers, positive real numbers.
- $[x,y]$: the segment joining two points $x,t \in \mathbb{R}^n$, which means the set of all points of the form $\lambda x + (1 - \lambda)y$ for some $0 \leq \lambda \leq 1$. In particular, if $n = 1$, this is just the familiar *interval* $[x,y]$. If we wish to exclude (for instance) y, we write $[x,y)$.

Terminology, Abbreviations, and Typographical Conventions

- $\nabla f(x)$: the gradient of a scalar-valued map (in this text we never have to use the derivatives of a vector-valued map) $f : \mathbb{R}^n \to \mathbb{R}$:

$$\nabla f(x)(i) := \frac{\partial f}{\partial x_i}(x).$$

- $\nabla^2 f(x)$: the Hessian matrix of a map $f : \mathbb{R}^n \to \mathbb{R}$:

$$\nabla^2 f(x)(i, j) := \frac{\partial^2 f}{\partial x_i \partial x_j}(x).$$

- $H(X), H(p)$: entropy of a random variable, of a probability vector
- $\Gamma(\alpha), B(\alpha, \beta)$: the Euler gamma and beta functions:

$$\Gamma(\alpha) := \int_0^\infty e^{-t} t^{\alpha-1}\, dt, \quad B(\alpha, \beta) := \int_0^1 t^{\alpha-1}(1-t)^{\beta-1}\, dt.$$

- $I_0(z)$: the modified Bessel function of order zero :

$$I_0(z) := \frac{1}{\pi} \int_0^\pi \exp\left(x \cos t\right) dt.$$

- $D(f), R(f)$: the domain and range of a map f. The notation $f : X \to Y$ means that the function f is defined on some subset of X, the domain $D(f)$. This means that what we call a function is what some people call a *partial function*. The raison d'être for this choice is that we use X just as a way to specify the nature of the variable (e.g., scalar or vector), without having to make assumptions about $D(f)$.
- $\mathcal{P}_n$: the set of n-component probability vectors:

$$\mathcal{P}_n := \{p \in \mathbb{R}^n : p_i \geq 0, p_1 + \cdots + p_n = 1\}.$$

- $\mathcal{P}_n^*$: the set of n-component probability vectors with positive components:

$$\mathcal{P}_n^* := \{p \in \mathbb{R}^n : p_i > 0, p_1 + \cdots + p_n = 1\}.$$

- $\log_2 x$: the logarithm to the base two; if a computation or statement is valid independently of the base, then the notation $\log$ is used. In this case, we assume only that the base is strictly greater than 1 (so that the corresponding logarithm function is increasing).
- $E_p(\phi)$: the expectation of ϕ under the probability distribution p. In other words, $E_p(\phi) := E(\phi(X))$, where X is a random variable having probability distribution p.
- $\mathcal{A}^+$: the set of all finite strings over a finite set $\mathcal{A}$.
- c^+: the extension to $\mathcal{A}^+$ of a binary code c defined on $\mathcal{A}$.
- $l_c(x)$: the code length of the symbol x, that is, the length of the codeword $c(x)$.

- $L(c)$: the average code length of the code c.
- $d_H(x, y)$: the Hamming distance between two binary words x and y of the same length.
- $\overset{d}{=}$: equality in distribution.
- $\overset{d}{\to}, \overset{p}{\to}, \overset{a.s.}{\to}$: convergence in distribution, in probability, almost sure.
- $\sim$: distributed as:
  1. $X \sim \mathcal{P}(\lambda)$ means that X is a Poisson variable of parameter λ;
  2. $X \sim \mathcal{B}(n, p)$ means that X is a binomial variable of parameters n and p;
  3. $X \sim \mathcal{NB}(k, p)$ means that X is a negative binomial variable of parameters k and p;
  4. $X \sim \mathcal{G}(p)$ means that X is a geometric variable of parameter p;
  5. $X \sim \mathcal{E}(\lambda)$ means that X is an exponential variable of parameter λ;
  6. $X \sim \mathcal{TE}(\lambda)$ means that X is a truncated exponential variable of parameters λ and n;
  7. $X \sim \Gamma(k, \lambda)$ means that X is a Γ variable of parameters λ and k.

Part I

Markov processes

1

Discrete time, countable space

Summary. Throughout this chapter, S is a countable space, called the *state space*; a *discrete-time stochastic process on* S is a collection X of $S-$valued random variables $(X_n)_{n \in \mathbb{N}}$ indexed by time n. We shall begin by formalizing the idea of lack of memory, which will be made rigorous in the definition of Markov processes. The use of the law of total probability will lead to the definition of the transition matrix, which encodes the dynamics of X. The problem is then to extract from this transition matrix information about the behavior of X. This problem is tackled in this chapter mostly by probabilistic methods, and we shall return to it with linear-algebraic tools in the next chapter. Although most results on the asymptotic behavior are quite intuitive, the proofs sometimes tend to be rather technical; readers interested mostly in the modeling aspects of Markov processes could limit themselves to the first five sections and then take a look at the statement of the main convergence result, Theorem 1.39; finally, some results specific to the case that S is finite are collected in the last section.

1.1 Conditional probability on a discrete space

Loosely speaking, a Markov process is a stochastic process whose probability distribution evolves in time in a memoryless fashion. Let us try to make things a little more precise in the case that time is discrete and see how the probability distribution at each time may be related to the probability distribution at previous times. Absolute lack of memory would mean that these two distributions have nothing to do with each other, or in mathematical language, that the corresponding variables are independent. Such a process is called *white noise*, and obviously it is useless if one tries to do any form of prediction; this memory loss requirement therefore needs to be weakened, and we will consider processes for which at every time, the probability distribution is entirely determined by what it was at previous times. The precise mathematical definition will be given in terms of conditional probabilities, so let's get started with a brief reminder on this notion.

© Springer International Publishing AG, part of Springer Nature 2018
J.-F. Collet, *Discrete Stochastic Processes and Applications,*
Universitext, https://doi.org/10.1007/978-3-319-74018-8_1

1 Discrete time, countable space

If A and B are two events from a probability space $(\Omega, \mathcal{F}, P)$ with $P(B) > 0$, the conditional probability $P(A|B)$ is defined as

$$P(A|B) := \frac{P(A \cap B)}{P(B)}.$$

You have seen in your basic probability class that for every such B, the map $A \mapsto P(A|B)$ is indeed a probability measure. In what follows we will make use of the following elementary facts:

Lemma 1.1. *Let $(\Omega, \mathcal{F}, P)$ be any probability space. For every finite collection of events $B_0, B_1, \ldots, B_n \in \mathcal{F}$,*

$$P(B_n \cap B_{n-1} \cap \cdots \cap B_1 | B_0)$$
$$= P(B_n | B_{n-1} \cap \cdots \cap B_1 \cap B_0) P(B_{n-1} | B_{n-2} \cap \cdots \cap B_1 \cap B_0) \cdots P(B_1 | B_0). \tag{1.1}$$

If, moreover, the sets $B_0, B_1, \ldots, B_n$ form a partition of Ω, then for every event A we have the so-called law of total probability:

$$P(A) = \sum_{i=0}^{n} P(A|B_i) P(B_i). \tag{1.2}$$

Finally, if B and C are two disjoint events, then for every event A we have

$$P(A|B \cup C) \leq \max\{P(A|B), P(A|C)\}. \tag{1.3}$$

Proof. The first relation is an immediate consequence of the definition of conditional probability (note that the right-hand side is a product of n fractions); the second relation simply expresses the fact that A is the disjoint union of the $A \cap B_i$'s. An upper bound for $P(A|B \cup C)$ is obtained as follows:

$$P(A|B \cup C) = \frac{P((A \cap B) \cup (A \cap C))}{P(B \cup C)} = \frac{P(A \cap B)}{P(B \cup C)} + \frac{P(A \cap C)}{P(B \cup C)}$$
$$= P(A|B) \frac{P(B)}{P(B \cup C)} + P(A|C) \frac{P(C)}{P(B \cup C)}$$
$$\leq \max\{P(A|B), P(A|C)\}.$$

$\square$

1.2 Formal definition of a Markov chain on a countable space

We now have the tools to formalize the idea of short-term memory described above.

Definition 1.2. *Let* $X := (X_n)_{n \in \mathbb{N}}$ *be a discrete-time stochastic process on* $\mathcal{S}$. *Choose a time* n *(think of it as the present), an integer* p, *and a collection of* p *times and* $p + 2$ *states:*

$$0 \le t_1 \le \cdots \le t_p < n, \quad x, y, y_1, \ldots, y_p \in \mathcal{S}.$$

The process X *is said to be a Markov process (or a Markov chain) if for every such choice we have*

$$P(X_{n+1} = y | X_n = x \cap X_{t_p} = y_p \cap \cdots \cap X_{t_1} = y_1)$$
$$= P(X_{n+1} = y | X_n = x).$$

If you think of n as the present time, the t_k's are p chosen times in the past; x is the current state, and the states y_k are prescribed positions in the past; let's call them "history"; then the definition basically says that the probability for the system to undergo a transition from x to y between times n and $n+1$ does not depend on the history.

As an extremely basic example consider a discrete-time game in which the player starts with an amount of $x \in \{1, 2, 3, 4\}$, and at each step earns 1 with probability 0.4, and loses 1 with probability 0.6. Let us decide that the game stops whenever the value 0 or 5 is reached, so that $\mathcal{S} = \{0, 1, 2, 3, 4, 5\}$. Calling X_n the amount at time n, we have

$$P(X_{n+1} = 5 | X_n = 5) = P(X_{n+1} = 0 | X_n = 0) = 1,$$

and for every $0 < i < 5$,

$$P(X_{n+1} = i + 1 | X_n = i) = 0.4, \quad P(X_{n+1} = i - 1 | X_n = i) = 0.6.$$

1.3 Homogeneous chains and transition matrices

The probability in the last definition depends only on x, y, and n, and it will be called the *probability of transition from* x *to* y *between times* n *and* $n + 1$. One further simplification consists in assuming that it does not depend on n; intuitively, this is like saying that the rule of the game does not change with time. In this case, the chain is said to be *homogeneous*:

Definition 1.3. *A Markov chain* $(X_n)_{n \in \mathbb{N}}$ *on* $\mathcal{S}$ *is said to be homogeneous if for every pair of states* $x, y \in \mathcal{S}$, *the transition probability* $P(X_{n+1} = y | X_n = x)$ *does not depend on* n:

$$\forall x, y \in \mathcal{S}, n \ge 0 : P(X_{n+1} = y | X_n = x) = P(X_1 = y | X_0 = x).$$

The dynamics of the system will be encoded in these transition probabilities, or more precisely in the map p defined on $\mathcal{S} \times \mathcal{S}$ by

$$p(x, y) := P(X_1 = y | X_0 = x).$$

1 Discrete time, countable space

If $S = \{x_1, \ldots, x_d\}$ is finite, then p may be identified with the $d \times d$ matrix $(P(X_1 = x_j | X_0 = x_i))_{1 \leq i,j \leq d}$; if S is infinite, this object becomes a doubly infinite matrix. In either case, by a slight abuse of terminology the map p will be called the *transition matrix* of the chain. We will, however, maintain the functional notation $p(x, y)$, which is standard in the literature on stochastic processes (unfortunately, some texts use a different convention and define $p(x, y)$ as the probability of going from y to x). We will therefore (get used to it) talk about the *row* x or *column* y of p, whereas x and y are elements of S.

In order to have consistent notation, we will also use a functional notation for probability distributions; thus, saying that q is the probability vector (whereas truly speaking we should call it the *probability mass function*) of a random variable X taking values in S amounts to writing

$$\forall x \in S : q(x) = P(X = x).$$

All entries of the transition matrix p being probabilities, they must lie between 0 and 1, and each row should be a probability vector; in other words, all row sums should be equal to 1. Let us formalize this:

Definition 1.4. *A Markov matrix, or stochastic matrix, on a countable set S is a map $p : S \times S \to [0, 1]$ satisfying the following condition:*

$$\forall x, y \in S : \sum_{y \in S} p(x, y) = 1.$$

1.4 The Chapman–Kolmogorov equation

The law of total probability gives a relationship between the probability vector of X_{n+1} and that of X_n:

$$P(X_{n+1} = x) = \sum_{y \in S} P(X_{n+1} = x \cap X_n = y) = \sum_{y \in S} p(y, x) P(X_n = y).$$

If we denote them respectively by q_{n+1} and q_n, this assumes the very compact form

$$q_{n+1} = q_n p. \tag{1.4}$$

This is the famous *Chapman–Kolmogorov equation* ("CK equation" from now on), which in words says that the probability vector at every time is obtained by multiplying the probability vector at the previous time by the transition matrix. Be aware that here q_n and q_{n+1} are row vectors, and in the product the matrix p is on the right (for textbooks using the transposed convention the probability vectors are column vectors, and p will be on the left). With our functional notation, this gives

$$\forall x, y \in \mathcal{S}: \; q_{n+1}(y) = \sum_{x \in \mathcal{S}} q_n(x)p(x,y).$$

This is very intuitive: the probability for the system to be in state y at time $n+1$ is a weighted sum of the probabilities for the system to be here or there at previous times, the weights being the transition probabilities. Iterating (1.4) we obtain

$$q_n = q_0 p^n, \quad \text{or} \quad q_{n+k} = q_n p^k \quad \forall k, n \geq 0.$$

This tells us what the k-step transition probability is:

$$P(X_{n+k} = y | X_n = x) = p^k(x,y).$$

Take care to not confuse q_n (the probability vector of X_n) with p^n (the pth power of the matrix p), nor $p^k(x,y)$ with $(p(x,y))^k$. The right-hand side of the last relation is the (x,y)-entry of the matrix p^k. A many-step transition may be split in two as we please, and in what follows we will make frequent use of the following immediate consequence of (1.4), as well as the lower bound that goes with it:

$$\forall x, y \in \mathcal{S}: \quad p^{n+m}(x,y) = \sum_{z \in \mathcal{S}} p^n(x,z)p^m(z,y), \tag{1.5}$$

$$\forall x, y, z \in \mathcal{S}: \quad p^{n+m}(x,y) \geq p^n(x,z)p^m(z,y). \tag{1.6}$$

As in any evolution problem, the issue of the asymptotic behavior arises: in loose terms, what does the vector q_n look like when n is very large? From (1.4) we see that if q_n has limit π as $n \to \infty$, we must necessarily have $\pi p = \pi$. This motivates the following definition:

Definition 1.5. *A probability vector π is said to be a stationary distribution if it satisfies $\pi p = \pi$.*

Of course in this case we will have $\pi p^n = \pi$ for every n. In components, the condition becomes

$$\sum_{x \in \mathcal{S}} \pi(x)p(x,y) = \pi(y), \quad \forall y \in \mathcal{S}. \tag{1.7}$$

Obviously, this last relation is automatically satisfied if π satisfies the following (more restrictive) condition:

Definition 1.6. *A probability vector π is said to be a detailed balance equilibrium if*

$$\pi(x)p(x,y) = \pi(y)p(y,x), \forall x, y \in \mathcal{S}. \tag{1.8}$$

If a detailed balanced equilibrium exists, the chain is said to be reversible.

Of course for the moment, this is all wishful thinking: a stationary distribution if present is a candidate for being a limit, but how can we establish that there is one, and that it indeed attracts q_n?

1.5 Transient and recurrent states

In light of the preceding section one might think of the study of the asymptotic behavior purely as a question in linear algebra (at least for the finite-space case): given a stochastic matrix p, how do its successive powers p^n behave? This point of view will be taken in next chapter, but for the moment we are to take a more probabilistic point of view. Roughly speaking, the approach will consist in studying the role that each state x may play in the large-time asymptotics. In order to do so, let us start by formalizing the idea of a state that is revisited infinitely many times:

Definition 1.7. *The first return time of a state y (this terminology is a bit unfortunate: unless the chain starts at y this is rather a first passage; but this is of no real consequence, since we will be interested mostly in the law of T_y conditioned on $\{X_0 = y\}$, in which case it is indeed a return) is the random variable T_y defined by*

$$T_y := \ min\{n \geq 1/X_n = y\},$$

with the usual convention that the minimum of the empty set is $+\infty$:

$$T_y := +\infty \quad if \quad X_n \neq y \quad \forall n \geq 1.$$

In other words, $T_y := +\infty$ means that the chain never visits the state y for $n \geq 1$.

The probability of this event will depend on the initialization, and we will make use of the following notation:

$$\rho_{x,y} := P(T_y < \infty | X_0 = x), \rho_{x,y}^n := P(T_y < n | X_0 = x).$$

To put words to this: starting from x, $\rho_{x,y}$ is the probability that y is visited in finite time, and $\rho_{x,y}^n$ is the probability that this occurs in less than n units of time.

The connection between the two is the following:

Lemma 1.8. *For every pair of states x, y, the numerical sequence $(\rho_{x,y}^n)_{n \in \mathbb{N}}$ increases to its limit $\rho_{x,y}$.*

Proof. This is an immediate consequence of the fact that the event $\{T_y < \infty\}$ may be represented as an increasing union:

$$\{T_y < \infty\} = \cup_{n \geq 1}\{T_y < n\}.$$

$\square$

We now have the tools to formalize the idea of a state that may be visited infinitely many times:

Definition 1.9. *A state y is said to be transient if $\rho_{y,y} < 1$, and is said to be recurrent if $\rho_{y,y} = 1$.*

If y is transient, there is a strictly positive probability $1 - \rho_{y,y}$ that the chain after starting at y never returns to it; on the other hand, if y is recurrent, it will be revisited in finite time with probability 1; intuitively, we should be able to iterate this to establish that the chain will in fact visit y infinitely many times; the rigorous proof will take quite a bit of work.

In order to get a feeling for what may happen, let us return to the game of the previous section, this time (to make things simpler) stopping at 4. Intuitively, one might expect the system to go to either 0 or 4; in other words, we suspect these two states to be recurrent and every other state to be transient. Here is how to show that (for instance) 1 is transient: if the chain starts at 1 and happens to visit 0, then it will never return to 1; therefore, the probability of never returning to 1 is bounded from below as follows:

$$P(T_1 = \infty | X_0 = 1) \geq p(1,0) = 0.6.$$

Hence we have

$$\rho_{1,1} = P(T_1 < \infty | X_0 = 1) = 1 - P(T_1 = \infty | X_0 = 1) \leq 0.4,$$

which indeed shows that 1 is transient. Obviously, the same reasoning applies to 2, considering the route $2 \to 1 \to 0$. More generally, a state x will be transient if either one of the two "black holes" may be reached from it. This motivates the following:

Definition 1.10. *If x and y are two states with $x \neq y$, then y is said to be accessible from x (which we will denote by $x \to y$) if $\rho_{x,y} \neq 0$. By convention, every state is accessible from itself. Two states x and y are said to communicate (which we will denote by $x \leftrightarrow y$) if each one is accessible from the other. A state x is said to be absorbing if $p(x,x) = 1$.*

In other words, y is accessible from x if and only if it is possible to go from x to y in a finite number of steps; of course, this property is encoded in the transition matrix (and is taken as a definition of accessibility in some texts):

Lemma 1.11. *State y is accessible from state x if and only if $p^n(x,y) > 0$ for some $n \geq 0$.*

Proof. For the case $x = y$ there is nothing to prove (take $n = 0$), so we may assume that $x \neq y$. The condition is obviously sufficient: in view of the inclusion of events

$$\{X_n = y\} \subset \{T_y \leq n\} \subset \{T_y < \infty\}$$

we have the inequality

$$P(X_n = y | X_0 = x) \leq P(T_y < \infty | X_0 = x),$$

which means $p^n(x,y) \leq \rho_{x,y}$. Let us show necessity arguing by contradiction: if $p^n(x,y) = 0$ for all n, then $\rho_{x,y}^n = 0$; thus from Lemma 1.8, we have $\rho_{x,y} = 0$. □

This makes it possible to show that the accessibility relation (and therefore the communication relation) is transitive by composing itineraries. More precisely, if $x \to y$ and $y \to z$, then we have $p^n(x,y) > 0$ and $p^m(y,z) > 0$ for some integers n, m. Thus

$$p^{n+m}(x,z) = \sum_{t \in \mathcal{S}} p^n(x,t) p^m(t,z) \geq p^n(x,y) p^m(y,z) > 0.$$

The communication relation is therefore an equivalence (the convention that every state is accessible from itself was in fact artificially added to ensure reflexivity), and later we will become interested in the equivalence classes. On the other hand, accessibility is clearly not symmetric (think, for instance, of an absorbing state). Let us see what happens when accessibility works in only one direction between two states:

Theorem 1.12. *Assume that two states x, y are such that y is accessible from x, but x is not accessible from y. Then x is transient.*

Proof. The idea is that starting from x it is possible to visit y, after which x will never be visited again. Let k be the smallest integer such that $p^k(x,y) \neq 0$, and assume that $X_0 = x, X_k = y$. Then (from the definition of k) for every $l < k$ we must have $X_l \neq x$, and in fact, for every $n > k$ we will also have $X_n \neq x$, since x is not accessible from y. Therefore, we get the inclusion of events

$$\{X_0 = x \cap X_k = y\} \subset \{X_0 = x \cap X_n \neq x \, \forall n \geq 1\},$$

from which we get

$$P(T_x = \infty | X_0 = x) \geq p^k(x,y) > 0,$$

which shows that x is transient. □

In fact, this idea may be pushed a bit further: if one may go from x to y but then has a nonzero probability of never returning to x, then x is transient:

Theorem 1.13. *Assume that two states x, y are such that $\rho_{xy} > 0$ and $\rho_{yx} < 1$. Then x is transient.*

The proof is just an adaptation of the previous one: take a path of n steps going from x to y such that y is reached at the nth step and never before, and then use it to bound $P(T_x = \infty | X_0 = x)$ from below.

Recurrence may sometimes be shown by deriving upper bounds from the first return time. Let us consider, for example, the following transition matrix on the space $\{1, 2, 3\}$:

$$p = \begin{pmatrix} 0.4 & 0.6 & 0 \\ 0.2 & 0.5 & 0.3 \\ 0.1 & 0.7 & 0.2 \end{pmatrix}.$$

Examination of the first column of p shows that for every state x, the probability of going from x to 1 is larger than 0.1:

$$P(X_{n+1} = 1 | X_n = x) \geq 0.1,$$

or equivalently,

$$P(X_{n+1} \neq 1 | X_n = x) \leq 0.9.$$

Using (1.1), we may then estimate the first return time as follows:

$$\begin{aligned} P(T_1 > n | X_0 = 1) &= P(X_n \neq 1 \cap X_{n-1} \neq 1 \cap \cdots \cap X_1 \neq 1 | X_0 = 1) \\ &= P(X_n \neq 1 | X_{n-1} \neq 1, \ldots, X_1 \neq 1, X_0 = 1) \\ &\quad P(X_{n-1} \neq 1 | X_{n-2} \neq 1 \cap \cdots \cap X_1 \neq 1 \cap X_0 = 1) \\ &\quad \cdots P(X_1 \neq 1 | X_0 = 1) \\ &= P(X_n \neq 1 | X_{n-1} \neq 1) P(X_{n-1} \neq 1 | X_{n-2} \neq 1) \\ &\quad \cdots P(X_1 \neq 1 | X_0 = 1). \end{aligned}$$

All probabilities in the last product may be bounded from above using (1.3), as for example,

$$\begin{aligned} P(X_n \neq 1 | X_{n-1} \neq 1) &= P(X_n \neq 1 | (X_{n-1} = 2) \cup (X_{n-1} = 3)) \\ &\leq \max\{P(X_n \neq 1 | X_{n-1} = 2), P(X_n \neq 1 | X_{n-1} = 3)\} \leq 0.9. \end{aligned}$$

Finally, we obtain the upper bound:

$$P(T_1 > n | X_0 = 1) \leq (0.9)^n.$$

Letting n go to infinity, we obtain

$$P(T_1 = \infty | X_0 = 1) = 0,$$

which shows that 1 is recurrent.

This tedious reasoning has no doubt convinced the reader of the need for simple criteria for deciding whether a state is recurrent or transient.

First of all some notation. For every state x we define the following quantities:

$$N(x) := \operatorname{card}\{n \geq 1 / X_n = x\}, \tag{1.9}$$

$$T_x^1 := T_x, \quad \forall k \geq 2: \quad T_x^k := \min\{n > T_x^{k-1} / X_n = x\}, \tag{1.10}$$

$$\forall k \in \mathbb{N}: \quad g_x^k := P(N(x) \geq k). \tag{1.11}$$

11

1 Discrete time, countable space

A few words of explanation: $N(x)$ and T_x^k are random variables, whereas g_x^k is a probability. The variable $N(x)$ is the (possibly infinite) number of visits to x; T_x^k is the time of the kth visit to x; we gave a "counting" definition for g_x^k as the probability for x to be visited at least k times, but we may reformulate it as a "waiting" definition:

$$g_x^k = P(T_x^k < \infty).$$

Let us begin with a few technical preliminaries:

Lemma 1.14. *For every state y,*

$$P(T_y^{k+1} < \infty | T_y^k < \infty) = \rho_{y,y}. \tag{1.12}$$

Proof. This result seems totally intuitive by shifting backward k units of time, but let us give a rigorous proof. From the inequalities $T_y^{k+1} \geq k+1$ and $T_y^{k+1} > T_y^k$ we get

$$P(T_y^{k+1} < \infty | T_y^k < \infty) = \frac{1}{P(T_y^k < \infty)} \sum_{l=k+1}^{\infty} \sum_{r=k}^{l-1} P(T_y^{k+1} = l \cap T_y^k = r). \tag{1.13}$$

From the Markov property we get

$$P(T_y^{k+1} = l \cap T_y^k = r) = P(T_y^{k+1} = l | T_y^k = r) P(T_y^k = r)$$
$$= P(X_l = y \cap X_{r+1} \neq y \cap \cdots \cap X_{l-1} \neq y | X_r = y) P(T_y^k = r)$$
$$= P(T_y^1 = l - r | X_0 = y) P(T_y^k = r),$$

so that returning to (1.13) yields

$$P(T_y^{k+1} < \infty | T_y^k < \infty)$$
$$= \frac{1}{P(T_y^k < \infty)} \sum_{l=k+1}^{\infty} \sum_{r=k}^{l-1} P(T_y^1 = l - r | X_0 = y) P(T_y^k = r). \tag{1.14}$$

Applying Fubini's theorem yields

$$P(T_y^{k+1} < \infty | T_y^k < \infty)$$
$$= \frac{1}{P(T_y^k < \infty)} \sum_{r=k}^{\infty} \sum_{l=r+1}^{\infty} P(T_y^1 = l - r | X_0 = y) P(T_y^k = r)$$
$$= \frac{1}{P(T_y^k < \infty)} \sum_{r=k}^{\infty} \sum_{j=1}^{\infty} P(T_y^1 = j | X_0 = y) P(T_y^k = r)$$
$$= P(T_y^1 < \infty | X_0 = y).$$

$\square$

Lemma 1.15. *For every* $k \geq 1$ *and pair of states* x, y, *we have*

$$P(T_y^k < \infty | X_0 = x) = \rho_{x,y}(\rho_{y,y})^{k-1}. \tag{1.15}$$

Proof. Intuitively, this result tautologically says that visiting y at least k times after having started at x amounts to going from x to y in finite time, and then $k - 1$ times from y to y. For $k = 1$ it is nothing but the very definition of $\rho_{x,y}$. Then in order to go from k to $k + 1$, we remark that from the Markov property we have

$$P(T_y^{k+1} < \infty | X_0 = x) = P(N(y) \geq k + 1 | X_0 = x)$$
$$= P(N(y) \geq k + 1 \cap N(y) \geq k | X_0 = x)$$
$$= P(N(y) \geq k + 1 | N(y) \geq k \cap X_0 = x) P(N(y) \geq k | X_0 = x)$$
$$= P(N(y) \geq k + 1 | N(y) \geq k) P(T_y^k < \infty | X_0 = x),$$

from which the conclusion follows thanks to (1.12). □

Lemma 1.16. *For every state* x, *the expected number of visits to* x *is given by*

$$E(N(x)) = \sum_{k=1}^{\infty} g_x^k. \tag{1.16}$$

Proof. From the definition of expectation we have

$$E(N(x)) = \sum_{k=1}^{\infty} k P(N(x) = k) = \sum_{k=0}^{\infty} k(g_x^k - g_x^{k+1}).$$

We may now sum by parts to get

$$k(g_x^k - g_x^{k+1}) = [k g_x^k - (k+1) g_x^{k+1}] + g_x^{k+1},$$

which yields the result. □

Lemma 1.17. *For every pair of states* x, y *we have (with the convention* $\frac{1}{0} = \infty$*)*

$$E(N(y) | X_0 = x) = \frac{\rho_{x,y}}{1 - \rho_{y,y}}. \tag{1.17}$$

Proof. The same computation as in the previous lemma gives

$$E(N(y) | X_0 = x) = \sum_{k=1}^{\infty} P(N(y) \geq k | X_0 = x),$$

and thus taking (1.15) into account, we obtain

$$E(N(y) | X_0 = x) = \sum_{k=1}^{\infty} \rho_{x,y}(\rho_{y,y})^{k-1}.$$

This series converges if and only if $\rho_{y,y} < 1$, that is, if and only if y is transient, in which case the result follows from the geometric series formula. □

Corollary 1.18. *If x is recurrent, then*

$$E(N(x)|X_0 = x) = +\infty.$$

Lemma 1.19. *For every pair (x, y),*

$$E(N(y)|X_0 = x) = \sum_{n=1}^{\infty} p^n(x, y). \qquad (1.18)$$

Proof. Noting that

$$E(\mathbb{1}_{X_n = y}|X_0 = x) = P(X_n = y|X_0 = x) = p^n(x, y),$$

we see that the number of visits to y is then evaluated as follows:

$$N(y) = \sum_{n=1}^{\infty} \mathbb{1}_{\{X_n = y\}}.$$

Taking expectations on both sides conditional on $X_0 = x$ gives the result. □

The last two lemmas give us a criterion for recurrence:

Proposition 1.20. *The state x is recurrent if and only if*

$$\sum_{n=1}^{\infty} p^n(x, x) = \infty.$$

Proof. If x is recurrent, then $\rho_{x,x} = 1$, and the result follows from (1.17) and (1.18); conversely, if the series diverges, then $p^n(x, x) \neq 0$ for some values of n, so $\rho_{x,x} \neq 0$; we may then use (1.17) and (1.18) to obtain $\rho_{x,x} = 1$. □

This criterion may then be used to show that recurrence is infectious:

Lemma 1.21. *If x is recurrent and $x \to y$, then y is recurrent.*

Proof. Necessarily x is accessible from y, for otherwise, from Theorem 1.12 x would be transient. Thus we have $\rho_{y,x} > 0$. Let us choose two integers j, l such that

$$p^j(y, x)p^l(x, y) \neq 0.$$

For every integer k we have the matrix relation $p^{j+k+l} = p^j p^k p^l$, from which we get

$$p^{j+k+l}(y, y) \geq p^j(y, x)p^k(x, x)p^l(x, y).$$

Summing over k, we get

$$\sum_{k=0}^{\infty} p^{j+k+l}(y, y) \geq p^j(y, x)p^l(x, y) \sum_{k=0}^{\infty} p^k(x, x).$$

This last sum is infinite, since x is recurrent. Therefore, from the same criterion, y also is recurrent. □

1.6 Hitting times

The first return time T_y defined in the previous section is the first time the chain enters, or "hits," the set $\{y\}$; more generally, we may be interested in the first time some property defined as $X_n \in A$ (where A is some subset of S) is satisfied:

Definition 1.22. *Let A be any subset of S; the hitting time of A is the random variable*

$$H_A := \min\{n \geq 0 / X_n \in A\},$$

and for every $x \in S$, the hitting probability of A starting from x is

$$h_A(x) := P(H_A < \infty | X_0 = x).$$

In words, $h_A(x)$ is the probability that starting from x, the chain reaches A in finite time; recall as before our convention that the minimum of the empty set is $+\infty$, meaning that $H_A = \infty$ if $X_n \notin A$ for all n. How the various quantities $h_A(x)$ are related when x changes is indicated by the following result:

Theorem 1.23. *The hitting probabilities satisfy the following:*

$$h_A(x) = \begin{cases} 1 & \text{for } x \in A, \\ \sum_{y \in S} h_A(y) p(x,y) & \text{for } x \notin A. \end{cases} \tag{1.19}$$

Proof. Note that the event that the chain reaches A in finite time may be written as the increasing union

$$\{H_A < \infty\} = \bigcup_{N=0}^{\infty} \{H_A \leq N\}.$$

By continuity of probability (Theorem B.1) this yields

$$h_A(x) = \lim_{N \to \infty} P(H_A \leq N | X_0 = x).$$

If $x \notin A$, then

$$P(H_A \leq N | X_0 = x) = P(\exists n \in \{1, \dots, N\} : X_n \in A | X_0 = x).$$

The idea is now to sum over all possible positions at some intermediate time, for instance time 1; using total probability, we obtain

$$P(H_A \leq N | X_0 = x)$$
$$= \sum_{y \in S} P(\exists n \in \{1, \dots, N\} : X_n \in A | X_1 = y \cap X_0 = x) P(X_1 = y | X_0 = x)$$
$$= \sum_{y \in S} P(\exists n \in \{1, \dots, N\} : X_n \in A | X_1 = y) p(x, y)$$
$$= \sum_{y \in S} P(\exists n \in \{0, \dots, N-1\} : X_n \in A | X_0 = y) p(x, y)$$
$$= \sum_{y \in S} P(H_A(y) \leq N - 1 | X_0 = y) p(x, y).$$

15

In the last sum each term is bounded by $p(x, y)$ and converges to $h_A(y)p(x, y)$ as $N \to \infty$; we may therefore use the dominated convergence theorem for series (a variant of Theorem A.8 in which N replaces the variable t) to conclude that the whole sum converges. $\square$

Remark: The statement given here is not optimal; depending on the nature of the transition matrix, the system (1.19) may or may not have a unique solution. If it is underdetermined, it will not suffice to determine the hitting probabilities; in this case, a sharper result (see, for instance, [26] Theorem 3.3.1, p. 112) characterizes h_A as the minimal nonnegative solution to (1.19). Our result will, however, suffice (coupled to a little extra work) to determine h_A in some simple birth-and-death or random-walk models (see Chapter 5) for which the transition matrix is sparse.

1.7 Closed sets and state space decomposition

Absorbing states if present will prevent all other states from being recurrent, since they trap the dynamics; a slightly more general situation is that in which a certain set of states plays the role of an absorbing state:

Definition 1.24. *A proper subset C of S is said to be closed if for every $x \in C$ and $y \notin C$, y is not accessible from x. The whole state space C is closed by convention. A closed set C is said to be irreducible if it contains no closed set other than itself.*

The entire space S may or may not be irreducible. It is easy to check that the intersection of two closed sets is also a closed set; for every $x \in S$ we may therefore consider the smallest closed set $\mathcal{O}_x$ containing x, which is the intersection of all closed sets containing x (this family is nonempty, since S is a member). Here is a simple characterization of irreducible closed sets:

Proposition 1.25. *A closed set C is irreducible if and only if all its elements communicate pairwise:*

$$\forall x, y \in C: \quad x \to y, \, y \to x.$$

Proof. The two-way accessibility condition is clearly sufficient; to show necessity, let us argue by contradiction and assume that we can find two states $x, y \in C$ such that y is accessible from x, but x is not accessible from y. Consider the set of points z that are accessible from y:

$$\mathcal{E} := \{z \in C: \quad y \to z\}.$$

From our assumption, $x \notin \mathcal{E}$, and it is easy to show that $\mathcal{E}$ is closed, which then implies that C is not irreducible (the reader is strongly advised to draw little circles and arrows pointing the right way). $\square$

For every $x \in S$, the *communicating class* of x is the set C_x of states y that communicate with x, or in other words, the equivalence class of x for the communication relation. Note that this class may or may not be a closed set, but if it is, from the previous proposition it is irreducible. We are now in a position to classify the states:

Theorem 1.26 (Classification of states, countable space). *The state space S may be written as a disjoint union*

$$S = C^1 \cup \cdots \cup C^r,$$

where $1 \le r \le \infty$, and the sets $C^1, \ldots, C^r$ are nonempty communicating classes. If a class is not closed, then it contains only transient states.

Beware! A closed class may contain transient states. We will see later that this never happens for finite chains, but it does happen, for instance, for some random walks.

Proof. The decomposition of S is just the partition induced by the communication equivalence; if a class C is not closed, pick two points $y \in C$ and $z \notin C$ with $y \to z$. Then y is not accessible from z, and from Theorem 1.12 we see that y is transient; now Lemma 1.21 tells us that every point in C also is transient. $\square$

1.8 Asymptotic behavior

1.8.1 Irreducibility

Definition 1.27. *A Markov chain is said to be* irreducible *if the whole space S is an irreducible closed set.*

Lemma 1.11 gives us an irreducibility criterion:

Proposition 1.28. *A chain is irreducible if and only if every transition is possible in finite time:*

$$\forall x, y \in S : \quad \exists n : p^n(x, y) > 0.$$

We are going to show that under an additional condition, an irreducible chain for which all states are recurrent admits a stationary distribution. Our first step towards this consists in showing that whatever the initial distribution, states are always visited infinitely many times:

Proposition 1.29. *Assume that a chain is irreducible and all states are recurrent. Then*

$$\forall x \in S : P(T_x < \infty) = 1.$$

1 Discrete time, countable space

Proof. From the law of total probability we have

$$P(T_x < \infty) = \sum_{y \in S} P(T_x < \infty | X_0 = y) P(X_0 = y),$$

so it is sufficient to show that $P(T_x < \infty | X_0 = y) = 1$ for all y, or equivalently that $P(T_x = \infty | X_0 = y) = 0$ for all y. Since every state y is recurrent, for every $m > 0$ we have

$$0 = P(T_y = \infty | X_0 = y)$$

$$= \sum_{z \in S} P(T_y = \infty | X_m = z \cap X_0 = y) P(X_m = z | X_0 = y)$$

$$= \sum_{z \in S} P(T_y = \infty | X_m = z) p^m(y, z)$$

$$= \sum_{z \in S} P(T_y = \infty | X_0 = z) p^m(y, z).$$

All terms in this last sum are zero, and in particular,

$$P(T_y = \infty | X_0 = x) p^m(y, x) = 0.$$

The chain is irreducible, so by Lemma 1.11 we may choose m to ensure that $p^m(y, x) \neq 0$, which yields the conclusion. $\square$

Let us now turn to the construction of a stationary distribution. With some initial state z being fixed, consider for every state x the number of passages by x before returning to z, and its expectation:

$$N_z(x) := \sum_{n=0}^{T_z - 1} \mathbb{1}_{\{X_n = x\}}, \quad \mu_z(x) := E(N_z(x) | X_0 = z). \tag{1.20}$$

Note that $\mu_z(z) = 1$, and that if x is not in the communicating class of z, then $\mu_z(x) = 0$. For the moment, z can be any state, and we will see later that it does not play much of a role. We require first a technical result about these quantities:

Lemma 1.30. *Assume that a chain is irreducible, and that all states are recurrent. Choose any state $z \in S$, and define the vector μ_z by (1.20); then for every pair of states $x, y \in S$ we have*

$$\mu_z(x) = \sum_{n=0}^{\infty} P(X_n = x \cap T_z > n | X_0 = z); \tag{1.21}$$

$$p(x, y) P(X_n = x \cap T_z > n | X_0 = z) = P(X_n = x \cap X_{n+1} = y \cap T_z > n | X_0 = z). \tag{1.22}$$

Proof. From proposition 1.29 we have $P(T_z < \infty) = 1$; therefore $N_z(x)$ may be computed by partitioning according to the values of T_z:

$$N_z(x) = \sum_{n=0}^{T_z-1} \mathbb{1}_{\{X_n=x\}} = \sum_{t=1}^{\infty} \mathbb{1}_{\{T_z=t\}} \sum_{n=0}^{T_z-1} \mathbb{1}_{\{X_n=x\}}$$

$$= \sum_{t=1}^{\infty} \sum_{n=0}^{T_z-1} \mathbb{1}_{\{X_n=x\cap T_z=t\}} = \sum_{t=1}^{\infty} \sum_{n=0}^{t-1} \mathbb{1}_{\{X_n=x\cap T_z=t\}}$$

$$= \sum_{n=0}^{\infty} \sum_{t=n+1}^{\infty} \mathbb{1}_{\{X_n=x\cap T_z=t\}}$$

(we have used Fubini's theorem for the last equality). Taking expectations, we get

$$E(N_z(x)|X_0 = z) = \sum_{n=0}^{\infty} \sum_{t=n+1}^{\infty} P(X_n = x \cap T_z = t|X_0 = z)$$

$$= \sum_{n=0}^{\infty} P(X_n = x \cap T_z > n|X_0 = z),$$

which establishes (1.21). For (1.22) we note that both terms of the equality vanish if $x = z$, so we need to treat only the case $x \neq z$. Let us transform the right-hand side:

$$P(X_n = x \cap X_{n+1} = y \cap T_z > n|X_0 = z)$$
$$= P(X_n = x \cap X_{n+1} = y \cap X_n \neq z \cap \cdots \cap X_1 \neq z|X_0 = z)$$
$$= P(X_n = x \cap X_{n+1} = y \cap X_{n-1} \neq z \cap \cdots \cap X_1 \neq z|X_0 = z)$$
$$= P(X_{n+1} = y|X_n = x \cap X_{n-1} \neq z \cap \cdots \cap X_1 \neq z \cap X_0 = z)$$
$$P(X_n = x \cap X_{n-1} \neq z \cap \cdots \cap X_1 \neq z|X_0 = z)$$
$$= p(x,y)P(X_n = x \cap X_n \neq z \cap X_{n-1} \neq z \cap \cdots \cap X_1 \neq z|X_0 = z)$$
$$= p(x,y)P(X_n = x \cap T > n|X_0 = z).$$

□

We can now show that the vector μ_z is an eigenvector of the transition matrix:

Lemma 1.31. *Under the same assumptions as in the previous lemma, for every $y \in S$ we have*

$$\mu_z(y) = \sum_{x \in S} \mu_z(x)p(x,y).$$

Proof. We first transform the right-hand side using (1.21) and (1.22):

$$\sum_{x \in S} \mu_z(x) p(x,y) = \sum_{x \in S} \sum_{n=0}^{\infty} P(X_n = x \cap T_z > n | X_0 = z) p(x,y)$$

$$= \sum_{n=0}^{\infty} \sum_{x \in S} P(X_n = x \cap T_z > n | X_0 = z) p(x,y)$$

$$= \sum_{n=0}^{\infty} \sum_{x \in S} P(X_n = x \cap X_{n+1} = y \cap T_z > n | X_0 = z)$$

$$= \sum_{n=0}^{\infty} P(X_{n+1} = y \cap T_z > n | X_0 = z)$$

$$= \sum_{n=1}^{\infty} P(X_n = y \cap T_z \geq n | X_0 = z).$$

Therefore, to conclude we just need to check the following equality:

$$\sum_{n=1}^{\infty} P(X_n = y \cap T_z \geq n | X_0 = z) = \sum_{n=0}^{\infty} P(X_n = y \cap T_z > n | X_0 = z).$$

If $y \neq z$, then for $n \geq 1$ both events $\{X_n = y \cap T_z \geq n\}$ and $\{X_n = y \cap T_z > n\}$ coincide, and the first term (corresponding to $n = 0$) in the sum on the right-hand side vanishes, so that the equality is satisfied. If $y = z$, the sum on the right-hand side reduces to the $n = 0$ term, which is 1; the sum on the left-hand side may be easily computed,

$$\sum_{n=1}^{\infty} P(X_n = z \cap T_z \geq n | X_0 = z) = \sum_{n=1}^{\infty} P(T_z = n | X_0 = z)$$

$$= P(T_z \geq 1 | X_0 = z) = 1,$$

so that the desired equality is again satisfied. $\square$

1.8.2 Positive recurrence

Having just obtained an eigenvector of the transition matrix, in order to get a stationary distribution we may wish to divide it by the sum of its components. This normalization would turn it into a probability vector, but a difficulty arises, due to the fact that the normalizing coefficient might be infinite. This quantity turns out to have a simple probabilistic interpretation: indeed, using (1.21), we have

$$\sum_{y \in S} \mu_z(y) = \sum_{y \in S} \sum_{n=0}^{\infty} P(X_n = y \cap T_z > n | X_0 = z) = \sum_{n=0}^{\infty} P(T_z > n | X_0 = z),$$

and therefore,

$$\sum_{y \in S} \mu_z(y) = E(T_z | X_0 = z). \tag{1.23}$$

This leads us to require that this quantity be finite; hence the following definition:

Definition 1.32. *The mean return time to a state z is the quantity $E(T_z | X_0 = z)$; the state z is said to be positive recurrent if its mean return time is finite, and the chain is said to be positive recurrent if all states are.*

Note that if a state y is transient, then its mean return time is infinite, since $P(T_y = \infty | X_0 = y) > 0$; this means that every positive recurrent state is recurrent, so our terminology is consistent. The recurrent states that are not positive recurrent are called *null recurrent states*; they are in a somewhat intermediate position between transient states and positive recurrent states. Normalization of the vector μ immediately gives us the following existence result:

Theorem 1.33. *Assume that a chain is irreducible and that all states are recurrent. Let z be a positive recurrent state, and for all $x \in S$ define*

$$\pi_z(x) := \frac{\mu_z(x)}{E(T_z | X_0 = z)},$$

where as before, $\mu_z(x)$ is defined by (1.20). Then the vector π_z is a stationary distribution.

Remark: using linear algebra, one may show that there cannot be more than one stationary distribution; for our construction in particular this implies that our π does not depend on z. In the case that the chain is ergodic (see below), we shall see in the next section that the uniqueness of the stationary distribution is a consequence of a convergence result, and that the expression of π may be simplified.

In fact, Theorem 1.33 may be slightly improved and turned into a characterization of positive recurrent chains:

Theorem 1.34. *For an irreducible chain, the following three conditions are equivalent:*

1. *There exists a positive recurrent state;*
2. *there exists a stationary distribution;*
3. *all states are positive recurrent.*

$1 \Rightarrow 2$ is just Theorem 1.33; for a proof of $2 \Rightarrow 3$ see, for instance, [26], p. 37. An irreducible chain is therefore positive recurrent, null recurrent, or transient. The strategy for determining which is the case will be first to look for a stationary distribution; if there is none, we are left with the transient/null recurrent alternative, which will be resolved by determining the hitting probabilities, making use of (1.19) (see Chapter 5 for examples). The following fact will come in handy:

Proposition 1.35. *Assume that a chain is irreducible. If there exists a pair of states* x, y *such that*

$$P(T_x < \infty | X_0 = y) < 1,$$

then the chain is transient.

This is an immediate consequence of Theorem 1.13. Indeed, since the chain is irreducible, x communicates with y, and by Theorem 1.13, x is transient. Transience being a class property, this means that the whole chain is transient. □

1.8.3 Periodicity and ergodicity

Under the assumptions of Theorem 1.33 we would like to see what else is needed to guarantee convergence to a stationary distribution; let us begin with a prototype of the situation whereby convergence of the vector p_n fails to occur. On the two-point space $\{1, 2\}$ consider the transition matrix p given by

$$p(1,1) = 0, \; p(1,2) = 1, \; p(2,1) = 1, \; p(2,2) = 0.$$

If we start the chain with a deterministic state (say, $P(X_0 = 1) = 1$), then the numerical sequence $P(X_n = 1)_{n \in \mathbb{N}}$ constantly alternates between 0 and 1. Let us note that this chain is irreducible but not ergodic: the possibility of an $n-$step transition depends on the parity of n. This type of periodic behavior is in fact the main obstruction to convergence; to investigate this further, let us define the period of a state:

Definition 1.36. *The period* d_x *of a state* x *is the greatest common denominator of the set of integers*

$$I_x := \{n \geq 1/p^n(x, x) > 0\};$$

the state x *is said to be aperiodic if* $d_x = 1$, *and the chain is said to be aperiodic if all states are aperiodic.*

Note that if $p(x, x) > 0$, then x is aperiodic, since $1 \in I_x$. The converse is obviously not true: x is aperiodic as soon as I_x contains two relatively prime integers; we will often make use of the following triviality:

$$p^n(x, x) > 0 \Rightarrow d_x \mid n.$$

Let us begin by showing that the period is a class property:

Theorem 1.37. *Two communicating states have the same period.*

In particular, if a chain is irreducible, then all states must have the same period.

Proof. Let x, y be two communicating states; call d_x, d_y their periods, and let l, m be two integers such that

$$\delta := p^l(x, y)p^m(y, x) > 0.$$

From (1.6) we have the lower bound

$$p^{l+m}(x, x) \geq p^l(x, y)p^m(y, x) > 0,$$

and thus d_x divides $l + m$. Using (1.6) again, we obtain

$$p^{l+d_y+m}(x, x) \geq p^l(x, y)p^{d_y}(y, y)p^m(y, x) = \delta p^{d_y}(y, y) > 0.$$

Therefore, d_x divides $l + d_y + m$, and so has to divide d_y. We may now swap x and y to conclude that d_y divides d_x, so finally, $d_y = d_x$. □

Aperiodicity will turn out to be the additional ingredient necessary for convergence; but first some vocabulary:

Definition 1.38. *A state is said to be ergodic if it is both aperiodic and positive recurrent; a chain is said to be ergodic if all states are.*

Let us now give the convergence result:

Theorem 1.39. *Assume that a chain is ergodic. Then it admits a unique stationary distribution π given by*

$$\forall x \in \mathcal{S}: \quad \pi(x) = \frac{1}{E(T_x \mid X_0 = x)},$$

and

$$\forall x \in \mathcal{S}: \quad \lim_{n \to \infty} \sum_{y \in \mathcal{S}} |p^n(x, y) - \pi(y)| = 0.$$

In particular, for every pair of states x, y we have the convergence

$$\lim_{n \to \infty} p^n(x, y) = \pi(y).$$

Proof. From Theorem 1.33 we know that a stationary distribution π exists. To show convergence, we use the so-called coupling method, which consists in defining an appropriate Markov process on the space $\mathcal{S}^2$, both components of which evolve (independently) according to the matrix p. To make things precise, define the $d^2 \times d^2$ square matrix q by setting

$$q((x_1, y_2), (x_2, y_2)) := p(x_1, y_1)p(x_2, y_2)$$

for every quadruple of states $x_1, x_2, y_1, y_2 \in \mathcal{S}$. It is immediate to check that q is indeed a Markov matrix. Now let (X_n, Y_n) be a Markov process on $\mathcal{S}^2$ evolving according to the transition matrix q, initialized with a pair X_0, Y_0 of independent $\mathcal{S}$-valued random variables. It is straightforward to check that

1 Discrete time, countable space

X_n and Y_n are independent for every $n \geq 0$. The *coalescence time* T is the random variable defined as the first time when X_n and Y_n coincide:

$$T := \min\{n \geq 0 / X_n = Y_n\}.$$

We are going to prove three technical lemmas about the variable T:

Lemma 1.40.
$$P(T < \infty) = 1. \tag{1.24}$$

Lemma 1.41. *For every state $y \in S$ and integer $n \geq 0$,*

$$P(X_n = y \cap T \leq n) = P(Y_n = y \cap T \leq n). \tag{1.25}$$

Lemma 1.42. *For every integer $n \geq 0$,*

$$\sum_{y \in S} |P(X_n = y) - P(Y_n = y)| \leq 2P(T > n). \tag{1.26}$$

Admitting these lemmas for the moment, let us show how Theorem 1.39 is proved. Take X_0 to be the constant x, and let Y_0 be distributed according to π:

$$P(X_0 = x) = 1, \quad \forall y : P(Y_0 = y) = \pi(y).$$

Inequality (1.26) becomes

$$\sum_{y \in S} |p^n(x, y) - \pi(y)| \leq 2P(T > n).$$

From (1.24) it follows that this upper bound converges to 0 as $n \to \infty$; thus each term on the left-hand side also converges to zero. The uniqueness of the stationary distribution is now a consequence of this convergence: keeping the same variable X_0 and taking another stationary distribution α (assuming that there is one) for the distribution of Y_0, we see that the component $(X_n)_{n \in \mathbb{N}}$ remains the same, and from the convergence result we get $\alpha = \mu$. Therefore, the distribution π_z provided by Theorem 1.33 does not depend on z, and we have $\pi(x) = \pi_x(x)$, which upon remarking that $\mu_x(x) = 1$ gives the expression for π. $\square$

For the sake of completeness let us now prove (1.24), (1.25), and (1.26).

Proof of (1.24). For every state $x \in S$, denote by $T_{x,x}$ the first time when X_n and Y_n coincide at x:

$$T_{x,x} := \min\{n \geq 0 / X_n = Y_n = x\}.$$

Obviously $P(T \leq T_{x,x}) = 1$, and thus $P(T < \infty) \geq P(T_{x,x} < \infty)$. From Proposition 1.29 we see that it is sufficient to check that the chain $(X_n, Y_n)_{n \in \mathbb{N}}$ is irreducible. The reader will easily check that for every $m \in \mathbb{N}$ and pair of states $x_1, x_2, y_1, y_2 \in S$ we have

$$q^m((x_1, y_2), (x_2, y_2)) = p^m(x_1, y_1)p^m(x_2, y_2).$$

By Theorem 1.48 there exists an integer m such that this quantity is nonzero for every quadruple $x_1, x_2, y_1, y_2 \in S$. This shows that the chain $(X_n, Y_n)_{n \in \mathbb{N}}$ is irreducible (and even ergodic).

Proof of (1.25). The idea consists in noting the following equality:

$$P(X_m = x \cap T = m) = P(X_m = Y_m = x \cap T = m) = P(Y_m = x \cap T = m).$$

Hence the strategy is to write

$$P(X_n = y, T \leq n) = \sum_{m=0}^{n} P(X_n = y \cap T = m)$$

$$= \sum_{m=0}^{n} \sum_{x \in S} P(X_n = y \cap T = m \cap X_m = x)$$

$$= \sum_{m=0}^{n} \sum_{x \in S} P(X_n = y | T = m \cap X_m = x)P(T = m \cap X_m = x).$$

The conditional probability above may be transformed using the Markov property and the fact that X and Y are independent:

$$P(X_n = y | T = m \cap X_m = x) =$$
$$P(X_n = y | X_m = x \cap Y_m = x \cap X_{m-1} \neq Y_{m-1} \cap \cdots \cap Y_0 \neq X_0)$$
$$= P(X_n = y | X_m = x \cap Y_m = x) = P(X_n = y | X_m = x) = p^{n-m}(x, y).$$

Therefore we obtain

$$P(X_n = y \cap T \leq n) = \sum_{m=0}^{n} \sum_{x \in S} p^{n-m}(x, y)P(T = m \cap X_m = x)$$

$$= \sum_{m=0}^{n} \sum_{x \in S} p^{n-m}(x, y)P(T = m \cap Y_m = x)$$

$$= P(Y_n = y \cap T \leq n),$$

which completes the proof of (1.25).

Proof of (1.26). We first split the quantity $P(X_n = y)$, making use of (1.25):

$$P(X_n = y) = P(X_n = y \cap T \leq n) + P(X_n = y \cap T > n)$$
$$= P(Y_n = y \cap T \leq n) + P(X_n = y \cap T > n).$$

We then get the upper bound

1 Discrete time, countable space

$$P(X_n = y) \leq P(Y_n = y) + P(X_n = y \cap T > n),$$

that is,

$$P(X_n = y) - P(Y_n = y) \leq P(X_n = y \cap T > n).$$

Of course, we may get an analogous upper bound by exchanging X and Y. It follows that

$$|P(X_n = y) - P(Y_n = y)| \leq \max(P(X_n = y \cap T > n), P(Y_n = y \cap T > n))$$
$$\leq P(X_n = y \cap T > n) + P(Y_n = y \cap T > n).$$

The result follows on summation over y. $\square$

In view of Theorem 1.34, we obtain the following corollary:

Corollary 1.43. *If an irreducible aperiodic chain has a stationary distribution π, then this stationary distribution is unique, and all the conclusions of Theorem 1.39 hold.*

1.9 Finite state space

The dynamics of Markov chains on a finite state space are somewhat simpler than on an infinite countable space, in particular in regard to asymptotic behavior. Throughout this section the state space $\mathcal{S}$ is assumed to be finite. Let us begin with three results on recurrence; the first one shows that a stationary distribution, if known, makes it possible to recognize recurrent states:

Proposition 1.44. *Suppose a chain admits a stationary distribution π. Then every state y such that $\pi(y) > 0$ is recurrent.*

Proof. Let y be such a state; we saw previously that for every $x \in \mathcal{S}$,

$$E(N(y)|X_0 = x) = \sum_{n=1}^{\infty} p^n(x, y) = \frac{\rho_{x,y}}{1 - \rho_{y,y}}.$$

Multiplying by $\pi(x)$ and summing over $x \in \mathcal{S}$ yields

$$\sum_{x \in \mathcal{S}} \pi(x) E(N(y)|X_0 = x) = \sum_{x \in \mathcal{S}} \pi(x) \sum_{n=1}^{\infty} p^n(x, y)$$

$$= \sum_{n=1}^{\infty} [\sum_{x \in \mathcal{S}} p^n(x, y) \pi(x)] = \sum_{n=1}^{\infty} \pi(y) = +\infty.$$

Therefore, there must be at least one state x for which the quantity $E(N(y)|X_0 = x)$ is infinite; so we must have $\rho_{y,y} = 1$, which means that y is recurrent. $\square$

Proposition 1.45. *Every irreducible closed set C is made up entirely of recurrent states.*

Proof. Closedness says that if a chain is initiated in C, then it never exits C. More precisely, for every $x \in C$, we have

$$P(\sum_{y \in C} N(y) = +\infty | X_0 = x) = 1, \tag{1.27}$$

and therefore

$$E(\sum_{y \in C} N(y) | X_0 = x) = +\infty.$$

Since C is finite, there is at least one point $y \in C$ such that

$$E(N(y) | X_0 = x) = +\infty.$$

From (1.17) we deduce that $\rho_{y,y} = 1$, which means that y is recurrent. Since x is accessible from y, Lemma 1.21 shows that x also recurrent. $\square$

Note that in the course of the proof we used only the fact that C is finite; in other words, we showed that even for an infinite state space, every closed finite irreducible set is made up of recurrent states only. As an immediate corollary, every finite Markov chain always has at least one recurrent state. With the help of these remarks we obtain the following consequence of Theorem 1.26:

Theorem 1.46 (Classification of states, finite space). *The set S may be partitioned as follows:*

$$S = T \cup C^1 \cup \cdots \cup C^r,$$

where T is the (possibly empty) set of transient states, $r \geq 1$ is some integer, and the sets $C^1, \ldots, C^r$ are nonempty irreducible closed communicating classes, therefore all consisting of recurrent states.

Proposition 1.47. *Every recurrent state is positive recurrent.*

Proof. Let z be a recurrent state. From (1.23) we see that showing that z is positive recurrent amounts to showing that the quantity $\sum_{y \in S} \mu_z(y)$ is finite, which means (since the sum has finitely many terms) showing that each term is finite. Note that the sum runs only over y in the communicating class C_z of z (all other terms vanish). The chain restricted to C_z is irreducible and has no transient state; we may thus apply Lemma 1.31 to obtain

$$\forall y \in C_z : \quad \mu_z(y) = \sum_{x \in C_z} \mu_z(x) p(x, y).$$

For $y \in C_z$, let k be an integer such that (we are using Lemma 1.11 here) $p^k(y, z) > 0$. Using the last relation with p^k in place of p, we obtain

1 Discrete time, countable space

$$1 = \mu_z(z) = \sum_{x \in C_z} \mu_z(x)p^k(x,z) \geq \mu_z(y)p^k(y,z),$$

from which we get the upper bound

$$\mu_z(y) \leq \frac{1}{p^k(y,z)} < \infty.$$

□

Combining the last two propositions, we see that if a chain is irreducible, then every state is positive recurrent; this means that on a finite space, every irreducible aperiodic chain is ergodic (in other words, Theorem 1.39 applies).

Let us close this section with a characterization of ergodicity that is specific to the case of a finite state space:

Theorem 1.48. *A chain is ergodic if and only if it satisfies the following condition for some integer n:*

$$\exists n : \forall x, y \in \mathcal{S} : p^n(x,y) > 0. \tag{1.28}$$

In other words, if a chain is ergodic on a finite space, then the integer n in Lemma 1.11 may be chosen uniformly with respect to the pair (x,y). Note that in this case, we will also have $p^m(x,y) > 0$ for every $m > n$. In some textbooks you will see that for finite Markov chains the condition (1.28) is taken as a definition of ergodicity.

Proof. Let us begin with sufficiency. If (1.28) holds, then obviously the chain is irreducible. Thus as remarked above all states are positive recurrent. We now need to show aperiodicity. For every x, the set

$$W_x := \{n \geq 1 : p^n(x,x) > 0\}$$

has a finite complement in $\mathbb{N}$; therefore, it has to contain infinitely many prime numbers. This means that the greatest common denominator of W_x is 1. In other words, x has period 1. Let us now show necessity, starting with the diagonal elements $p^n(x,x)$. First note that since $\mathcal{S}$ is finite, it suffices to show that for every x, the set W_x has a finite complement in $\mathbb{N}$. By aperiodicity, the elements of W_x are relatively prime. Let us now show that W_x is closed under addition. If we take any pair of integers $n, m \in W_x$, from (1.6) we have

$$p^{n+m}(x,x) \geq p^n(x,x)p^m(x,x) > 0.$$

Therefore, $n + m \in W_x$. The conclusion now follows from an elementary number-theoretic result that asserts that every set of relatively prime positive integers that is closed under addition has a finite complement in $\mathbb{N}$ (see Lemma D.1 in Appendix D). We may now treat the nondiagonal terms $p^n(x,y)$ by again using (1.6). □

1.10 Problems

1.1. Show that the product of two Markov matrices (of the same size) is also a Markov matrix. What sort of linear combinations of Markov matrices give Markov matrices?

1.2. On the two-point set $\{1, 2\}$ consider the Markov matrix defined by $P(1, 2) = p, P(2, 1) = q$. Let (X_n) be a Markov chain evolving according to p, with $P(X_0 = 1) = 1$. Compute the quantity $P(X_n = 1)$, and find its limit as $n \to \infty$.

1.3. A fair die is thrown n times, and we call X_n the largest number obtained (or more mathematically, we give ourselves n independent and identically distributed (i.i.d.) random variables all uniformly distributed on $\{1, \ldots, 6\}$ and define X_n to be the maximum).

1. Show that this defines a Markov chain on some state space $\mathcal{S}$ to be specified.
2. Give the transition matrix p.
3. Show that
$$p^{n+1}(i, 6) - p^n(i, 6) = \frac{1}{6}(1 - p^n(i, 6)).$$
4. Deduce that $p^n(i, 6)$ converges to 1 as n becomes large.
5. Does the chain have a stationary distribution?

1.4. Let (X_n) be a Markov chain on $\{0, 1\}$. If p is the transition matrix, set $p := p(1, 0)$ and $q := p(0, 0)$. Let T_1 be the first return time to 1.

1. Give expressions for the quantities $P(T_1 = 1 | X_0 = 1)$ and $P(T_1 \geq 2 | X_0 = 1)$ in terms of p and q.
2. The same question for the quantity $P(T_1 \geq n | X_0 = 1)$, for all n; be careful to pinpoint exactly how the Markov property comes into play.
3. Deduce the expression of $P(T_1 = n | X_0 = 1)$ in terms of p and q.

1.5. Let (X_n) be a finite Markov chain, and denote by p the transition matrix. Let x be any fixed state, and let T be the first positive time when the state of the chain is not x:
$$T := min\{n > 0 / X_n \neq x\}.$$

1. Show that $P(T > n | X_0 = x) = p(x, x)^n$.
2. Deduce the value of $P(T = n | X_0 = x)$.

1.6. There are N light bulbs, all of which are turned off at time 0; at every (discrete) time we select a bulb uniformly and turn it on if it is off, while leaving it lit if it is lit. Let X_n denote the number of lit bulbs at time n. Give the state space and transition matrix, and write down the CK (Chapman–Kolmogorov) equation. What can you say about the asymptotic regime?

1 Discrete time, countable space

1.7. Define a Markov chain (X_n) on the set $\{1, \ldots 5\}$ by the following rule: given the value of X_n, if it is less than 5, then the number X_{n+1} is drawn uniformly between $X_n + 1$ and 5; if $X_n = 5$, we set $X_{n+1} = 5$. Call T the first time at which 5 is reached.

1. Give the transition matrix p.
2. Give the set T of all possible values for T, and then find $P(T = k | X_0 = 1)$ for each $k \in T$ and compute the conditional expectation $E(T | X_0 = 1)$.
3. We now give ourselves the same rule, but on the set $\{1, \ldots, N + 1\}$. For every integer j between 1 and N, denote by N_j the number of passages by state j, and for every $k \in \mathbb{N}$ denote by X_j^k the Bernoulli variable that is 1 if $X_k = j$, and 0 otherwise. Give an expression for N_j in terms of the variables $(X_j^k)_k$, and deduce the relation

$$E(N_j | X_0 = 1) = \delta_{1,j} + p(1, j) + \cdots + p^n(1, j) + \cdots.$$

4. The transition matrix p is an $(N + 1) \times (N + 1)$ square matrix. Denote by Q the $N \times N$ block obtained from p by erasing its last row and last column. Show that $E(N_j) = N(1, j)$, where the matrix N is defined by $N := (I - Q)^{-1}$.
5. What are the recurrent states of this chain, and what is the meaning of the quantity $\sum_{k=1}^N N(1, k)$?

1.8. A virus may be present in one of N different forms, and it may switch from one to another, which is called a mutation. At every (discrete) time point, the virus undergoes a mutation with probability α, and when it does, it chooses uniformly from among the $N - 1$ other forms.

1. Use the law of total probability to find the transition matrix.
2. Find the stationary distribution, and show by a direct method that it is unique.

1.9. A box contains N balls, some white, some black. We draw a ball uniformly, and regardless of its color we throw it away; then with probability p we replace it with a white ball, with probability $1 - p$ with a black ball (we can do this because next to our box we have an infinite supply of balls). Note that the total number in our box never changes. Call X_n the number of white balls in the box after n operations.

1. Give the transition matrix (beware of the boundary).
2. For the case $N = 2$ show that the system has a unique stationary distribution, given by
$$\pi = ((1 - p)^2, 2p(1 - p), p^2).$$
3. Based on the previous question, can you find the stationary distribution for the general case?

1.10. Let N and p be two fixed positive integers. A box (not an urn) contains N balls, some red, some black. A ball is drawn uniformly, and its color is looked at. It is then put back in the box together with p additional balls of the same color, taken from an infinite supply of red and black balls. When we have done this once, we have a total of $N + p$ balls in the box. We keep going in the same manner, and we denote by X_n the number of red balls in the box after we have done n operations. Give the state space and the transition matrix.

1.11. Let us take a look at a classical diffusion model called the Ehrenfest urn. We have d particles, distributed between two boxes. At each time point we uniformly draw one of the d particles (meaning that we do not know out of which box it is going to be drawn) and then see which box it comes from and put it in the other box. Let X_n be the number of particles present in the left box after the nth operation.

1. Give the transition matrix p.
2. Justify without computation why $p^n(x, x) = 0$ if n is odd.

1.12. Now for a variant of the Ehrenfest urn. Taking the same setup as in the previous problem, we slightly change the rule of the game: after the particle has been drawn, we put it back where it came from with probability $\frac{1}{2}$ and put it in the other box with probability $\frac{1}{2}$; X_n is defined as before.

1. Give the state space and transition matrix.
2. Write down the CK equation, taking care to treat the first and last relations separately.
3. We are now going to look for a stationary distribution π. Show that necessarily π satisfies

$$\pi_0 = \frac{1}{d}\pi_1, \qquad \pi_d = \frac{1}{d}\pi_{d-1}, \qquad (1.29)$$

$$1 \le k \le d-1: \quad \pi_k = (1 - \frac{k-1}{d})\pi_{k-1} + (\frac{k+1}{d})\pi_{k+1}. \qquad (1.30)$$

4. Check that the sequence $a_k = \binom{d}{k}$ solves this recurrence relation, and deduce the components of π.

1.13. Consider binary sequences of infinite length, each bit generated with the same Bernoulli distribution and independently of all others. If bit number n is a 1, we set X_n equal to the length of the last block of 1's before position n; otherwise, we set $X_n = 0$. For example,

$$111110101010111\ldots \longrightarrow X_{15} = 3, \quad X_4 = 4, \quad X_6 = 0.$$

1. What are the possible transitions from the configuration $X_n = k$? Explain why $(X_n)_{n \in \mathbb{N}}$ is a Markov chain on $\mathbb{N}$.
2. Write down the CK equation, taking care in treating the case $k = 0$.

3. Deduce the existence of a stationary distribution.

1.14. This problem is about another proof of the convergence result (this is taken from the excellent monograph [36], which we recommend you take a look at). Let p be a Markov matrix on the space $\{1, \ldots, n\}$ (to lighten the notation, all minima, maxima, and sums here are understood on this set). Define the following quantities:

$$m_n(j) := \min_{i \in S} p_n(i, j), \quad M_n(j) := \max_{i \in S} p_n(i, j), \quad \delta := \min_{i,j \in S} p(i, j),$$

and assume that $\delta < 1$.

1. Show that for every fixed j, the sequence $(m_n(j))_{n \in \mathbb{N}}$ is increasing, and the sequence $(M_n(j))_{n \in \mathbb{N}}$ is decreasing.
2. Deduce that they are both convergent sequences.
3. For every pair i, j, show that

$$p_{n+1}(i, j) = \sum_{k \in S} p_n(k, j)[p(i, k) - \delta p_n(j, k)] + \delta p_{2n}(j, j).$$

4. Deduce that

$$m_{n+1}(j) \geq (1 - \delta)m_n(j) + \delta p_{2n}(j, j).$$

5. Using the same method, show that

$$M_{n+1}(j) \leq (1 - \delta)M_n(j) + \delta p_{2n}(j, j).$$

6. Deduce that for every j, the sequences $(m_n(j))_{n \in \mathbb{N}}$ and $(M_n(j))_{n \in \mathbb{N}}$ have the same limit as $n \to \infty$. We will denote it by π_j.
7. Show that the vector π is a stationary distribution.
8. Show that all states are recurrent.
9. Finally, show that convergence is geometric, i.e., show that

$$|p_n(i, j) - \pi_j| \leq (1 - \delta)^n.$$

Linear algebra and search engines

Summary. The first five sections of this chapter develop the machinery necessary to give a purely linear-algebraic proof of the convergence theorem for ergodic chains on finite spaces. It is probably fair to say that the key property of the matrices (apart from being stochastic matrices) is positivity; this is why in order to make the presentation self-contained, most classical results on the Perron–Frobenius theory are included. The last section deals with the by now classical PageRank algorithm, which is included here because it is traditional to present it in the language of linear algebra rather than that of Markov processes.

2.1 Spectra of Markov matrices

Recall that for a square matrix $A \in \mathcal{M}_n(K)$, the number $\lambda \in K$ is said to be an *eigenvalue* of A if there exists a nonzero column vector $x \in K^n$ such that $Ax = \lambda x$. If a nonzero row vector $y \in K^n$ satisfies $xA = \lambda x$, we say that x is a *row eigenvector* (also called *left eigenvector* in some texts) of the matrix A. In other words, x is an eigenvector of A if and only if x^t is a row eigenvector of A^t. The set of all eigenvalues of the matrix A is called the *spectrum* of A, denoted by $\sigma(A)$. If λ is an eigenvalue of A, the set of all vectors x satisfying $Ax = \lambda x$ (i.e., all associated eigenvectors plus the zero vector) is called the *eigenspace* associated with λ. This is obviously a linear space, and its dimension is called the *geometric multiplicity* of the eigenvalue λ. As you probably know, a number $\lambda \in K$ is an eigenvalue of A if and only if it is a root of the so-called *characteristic polynomial* of A, which is the polynomial p_A defined by

$$p_A(X) := \det(XI_n - A).$$

The multiplicity of the root λ is called the *algebraic multiplicity* of the eigenvalue λ. It can be shown that the geometric multiplicity is always less than or equal to the algebraic multiplicity. Recall that an eigenvalue λ is said to be *simple* if its algebraic multiplicity is 1; thus in this case, its geometric multiplicity also is 1.

© Springer International Publishing AG, part of Springer Nature 2018
J.-F. Collet, *Discrete Stochastic Processes and Applications,*
Universitext, https://doi.org/10.1007/978-3-319-74018-8_2

The *spectral radius* of A is the maximum modulus of all eigenvalues:

$$\rho(A) := \max\{|\lambda|, \lambda \in \sigma(A)\}.$$

For some matrices, $\rho(A)$ happens to be an eigenvalue; in this case and if no other eigenvalue has modulus $\rho(A)$, we shall say that $\rho(A)$ is the *dominant eigenvalue*. We shall see (see Theorem 2.15 below) that this happens, for instance, for positive matrices.

Most results on location of the eigenvalues are proved using what we might call the *maximum modulus component trick*, which may be described as follows. If x is an eigenvector of the matrix A associated with the eigenvalue λ, let i be an index such that x_i has maximum modulus, which means that for every other k we have $|x_k| \le |x_i|$ (note that in this case, since $x \ne 0$, we must have $x_i \ne 0$). Then the relation

$$\sum_{j=1}^{n} a_{ij} x_j = \lambda x_i \tag{2.1}$$

may be used to derive various inequalities for λ. The first result below on the spectrum of Markov matrices is perhaps one of the simplest illustrations of this method:

Theorem 2.1. *For every Markov matrix A we have $\rho(A) = 1$, and 1 is an eigenvalue.*

Proof. The fact that the column vector $x := (1, \dots, 1)^t$ (all components equal to 1) satisfies $Ax = x$ shows that 1 is an eigenvalue, and therefore $\rho(A) \ge 1$.

We shall now show that $\rho(A) \le 1$ using the maximum modulus component trick described above. Given some eigenvalue $\lambda \in \sigma(A)$, let $x \ne 0$ satisfy $Ax = \lambda x$, and let i be such that $|x_i| = \max_k |x_k|$. From (2.1) we get

$$|\lambda||x_i| \le \sum_{j=1}^{n} a_{ij}|x_i| = |x_i|,$$

and hence $\lambda \le 1$. Since this is true of every $\lambda \in \sigma(A)$, we obtain $\rho(A) \le 1$. Thus $\rho(A) = 1$. $\square$

Now that we know that the spectral radius is 1 and that 1 is indeed an eigenvalue of maximum modulus, two questions arise: can there be other eigenvalues of maximum modulus, and in the affirmative case, what can they be? This is the object of the next two results:

Theorem 2.2. *Let A be a Markov matrix. If all diagonal elements of A are nonzero, then the only eigenvalue of modulus 1 is 1.*

Proof. Maximum modulus component trick again: let $\lambda = e^{i\mu}$ be an eigenvalue of modulus 1, x an associated eigenvector, and x_i a maximum-modulus component. Subtracting $a_{ii}x_i$ from both sides of (2.1) and dividing by x_i yields

$$\lambda - a_{ii} = \sum_{j=1, j \neq i}^{n} a_{ij} \frac{x_j}{x_i}.$$

Taking moduli, we get

$$|\lambda - a_{ii}| \leq \sum_{j=1, j \neq i}^{n} a_{ij} = a_{ii} - 1,$$

and taking squares on both sides of this inequality, we obtain

$$a_{ii} \cos \mu \geq a_{ii}.$$

If $a_{ii} \neq 0$, we obtain $\cos \mu = 1$, which means precisely that $\lambda = 1$. $\square$

Theorem 2.3. *Let A be a Markov matrix, and $\lambda \in \sigma(A)$ an eigenvalue of modulus 1. Then λ is a complex root of unity: $\lambda^k = 1$ for some $k \in \mathbb{N}$.*

Proof. We may assume that $\lambda \neq 1$ (if $\lambda = 1$, there is nothing to prove). Let use choose x and i as in the proof of the previous theorem; since we have assumed $\lambda \neq 1$, we now know that $a_{ii} = 0$, and thus (2.1) becomes

$$\lambda = \sum_{j=1, j \neq i}^{n} a_{ij} \frac{x_j}{x_i}, \tag{2.2}$$

and taking the modulus, we obtain

$$1 = |\lambda| = | \sum_{j=1, j \neq i}^{n} a_{ij} \frac{x_j}{x_i}| \leq \sum_{j=1, j \neq i}^{n} a_{ij} |\frac{x_j}{x_i}| \leq \sum_{j=1, j \neq i}^{n} a_{ij} = 1. \tag{2.3}$$

Comparing the last two sums in this string of inequalities, we see that for every j such that $a_{ij} \neq 0$, we must have $|\frac{x_j}{x_i}| = 1$. Note that since $\sum_{j=1, j \neq i}^{n} a_{ij} = 1$, the set of such j's cannot be empty. Now comparing the first two sums in (2.3), we see that for all these integers j, the complex numbers $a_{ij} \frac{x_j}{x_i}$ must be collinear (since the modulus of their sum is the sum of their moduli). This means that for some $\alpha_{ij} \in \mathbb{R}$ and some $\theta \in \mathbb{R}$, we have

$$a_{ij} \frac{x_j}{x_i} = \alpha_{ij} e^{i\theta}.$$

Taking the modulus, we obtain $\alpha_{ij} = a_{ij}$, and thus $x_j = x_i e^{i\theta}$. Returning to (2.2) gives $\lambda = e^{i\theta}$, and thus $x_j = \lambda x_i$. To summarize, we have just shown that if we select a maximum-modulus component x_i of the eigenvector x, then some other component of x has to equal λx_i and is therefore a maximum-modulus component of x. Obviously we may iterate this reasoning to conclude that for every positive exponent p, the number $\lambda^p x_i$ is a maximum-modulus component of x. Since there are only finitely many such components, we see that the set of all positive powers of λ is finite, which means precisely that λ is a root of unity. $\square$

It is not widely known that the main spectral properties of Markov matrices that we just saw can be obtained in a much more elegant way from the following geometrical result due to Dmitriev and Dynkin (see [9, 10]):

Theorem 2.4 (Dmitriev–Dynkin theorem). *Let Λ_n denote the union of all spectra of all $n \times n$ stochastic matrices. Then a complex number λ is in Λ_n if and only if for some integer $q \leq n$ there exists a polygon $Q \subset \mathbb{C}$ with q vertices whose convex hull is closed under multiplication by λ: $\lambda Conv(Q) \subset Conv(Q)$.*

Proof. Let $\lambda \in \Lambda_n$ be an eigenvalue of a stochastic matrix A with corresponding eigenvector z:

$$\lambda z_j = a_{j,1} z_1 + \cdots + a_{j,n} z_n. \tag{2.4}$$

Let Q be the polygon with vertices $z_1, \ldots, z_n$ (note that since some components may be equal, the number q of such vertices may be strictly less than n). The relation above says precisely that $\lambda Q \subset Conv(Q)$, and therefore $\lambda Conv(Q) \subset Conv(Q)$. Conversely, if a polygon Q with q vertices satisfies this inclusion condition, then for every j less than q we may find q nonnegative coefficients $(a_{j,k})_{1 \leq k \leq q}$ such that (2.4) holds. This yields a stochastic $q \times q$ matrix A for which λ is an eigenvalue; if $q < n$, we may increase the size of A by adding zero entries, so as to get an $n \times n$ stochastic matrix. $\square$

Deceptively simple as it looks, this geometrical characterization of Λ_n may be used to prove Theorems 2.1 and 2.3 in a much simpler way. Let us start with Theorem 2.1 (recall that the only nontrivial part in this proof was the bound $\rho(A) \leq 1$). Let A be a stochastic matrix, $\lambda \in \sigma(A)$ some eigenvalue of A, and let the polygon Q be associated with λ as in Theorem 2.4. Pick a number $z \in Conv(Q)$ with maximum modulus. Since $\lambda z \in Conv(Q)$, we must have $|\lambda z| \leq |z|$. Therefore, $\lambda \leq 1$. $\square$

Let us now see how Theorem 2.3 is an easy corollary of Theorem 2.4. Taking p, λ, Q and z as in the previous proof, it is easy to see that z must be a vertex of Q. Then for every integer p we have $\lambda^p z \in Conv(Q)$, and $|\lambda^p z| = |z|$. This means that $\lambda^p z$ has maximum modulus in $Conv(Q)$, and so it must be a vertex of Q. Since there are only q vertices, this shows that the set of all the values $\{\lambda^p, p \in \mathbb{N}\}$ is finite, which means precisely that λ is a root of unity. $\square$

It is quite interesting to compare the first proof we saw above of Theorem 2.3 and this one, which clearly is shorter, more elegant, and easier to follow. In a way, the simplification and gain in elegance come from the fact that instead of looking at just one given stochastic matrix A (as was done in the first proof) we immerse it in the set of all stochastic matrices A and consider the geometry of the set Λ_n; somehow, it is this choice of a higher level of abstraction that makes things simpler. In the sense of Erdös, the proof of Theorem 2.3 from Theorem 2.4 probably is the *book proof.*

2.2 More on the location of eigenvalues for general square matrices

For an $n \times n$ matrix A, its *deleted row sums* and *deleted column sums* are the following quantities:

$$R_i(A) := \sum_{j=1,j\neq i}^{n} a_{ij}, \quad C_i(A) := \sum_{j=1,j\neq i}^{n} a_{ji}. \tag{2.5}$$

The following result tells how much the diagonal weighs against the rest of the matrix in determining where the eigenvalues may be:

Theorem 2.5 (Gershgorin's discs). *Let A be an $n \times n$ matrix. Then for every eigenvalue $\lambda \in \sigma(A)$ there exist an integer i such that*

$$|\lambda - a_{ii}| \le R_i(A)$$

and an integer j such that

$$|\lambda - a_{jj}| \le C_j(A).$$

The discs $G_i(A)$ defined in the complex plane by

$$G_i(A) := \{z \in \mathbb{C}: \quad |z - a_{ii}| \le R_i(A)\}$$

(and analogously with radius $C_i(A)$) are called the *Gershgorin discs of the matrix A;* thus the theorem may be restated as the inclusion

$$\sigma(A) \subset \bigcup_{i=1}^{n} G_i(A). \tag{2.6}$$

Proof. Since the deleted row sums of A are the deleted column sums of its transpose, it suffices to prove only the first half of the statement. Let x be an eigenvector associated with the eigenvalue λ, and choose i such that x_i is a maximum-modulus component of x. We have

$$\lambda x_i = \sum_{j=1}^{n} a_{ij} x_j,$$

and thus subtracting the term $a_{ii} x_i$, dividing by x_i, and taking the modulus, we obtain

$$|\lambda - a_{ii}| \le R_i(A).$$

$\square$

This result has an interesting corollary regarding Markov matrices:

Corollary 2.6. *Let A be a Markov matrix. If all diagonal elements of A are strictly larger than $\frac{1}{2}$, then A is invertible.*

Proof. Since A is a Markov matrix, we have $R_i(A) = 1 - a_{ii}$ for all i; thus if λ is an eigenvalue, we have

$$|\lambda - a_{ii}| \leq 1 - a_{ii}$$

for some i. Since the modulus function is Lipschitz continuous, this implies

$$||\lambda| - a_{ii}| \leq 1 - a_{ii},$$

and thus

$$|\lambda| \geq 2a_{ii} - 1.$$

This shows that $\lambda \neq 0$. Therefore, A is invertible.   □

Gershgorin's discs provide the best known example of an *inclusion set* for the spectrum of arbitrary square matrices; a slightly less classical example is given by Brauer's ovals:

Theorem 2.7 (Brauer's ovals). *Let A be an $n \times n$ matrix. Then for every eigenvalue $\lambda \in \sigma(A)$ we have the bound*

$$|\lambda - a_{ii}||\lambda - a_{jj}| \leq R_i(A)R_j(A)$$

for some pair of indices i, j with $i \neq j$.

The proof relies on the following lemma:

Lemma 2.8. *Let A be an $n \times n$ matrix. If either one of the two conditions*

$$\forall i, j, i \neq j : \quad |a_{ii}||a_{jj}| > R_i(A)R_j(A) \tag{2.7}$$

$$\forall i, j, i \neq j : \quad |a_{ii}||a_{jj}| > C_i(A)C_j(A) \tag{2.8}$$

is satisfied, then the matrix A is invertible.

Proof. Again (changing A into its transposed matrix) it suffices to prove the result for the condition on row sums, which we shall do by contradiction. Assuming that A is not invertible, let x be a (nonzero) null vector of A. We use the maximum component trick twice: first choose k_1 such that x_{k_1} is a maximum-modulus component of x, and let k_2 be a maximum-modulus component once x_{k_1} has been removed, meaning

$$\forall i \neq k_1 : \quad |x_i| \leq |x_{k_2}| \leq |x_{k_1}|.$$

Obviously, $x_{k_1} \neq 0$, since $x \neq 0$. If $x_{k_2} = 0$. Then x_{k_1} is the only nonzero component of x, and since $(Ax)_{k_1} = 0$, we obtain $a_{k_1 k_1} = 0$, and (2.7) is violated. If $x_{k_2} \neq 0$, using $(Ax)_{k_2} = 0$ we obtain the relation

$$a_{k_2 k_2} x_{k_2} = - \sum_{j=1, j \neq k_2}^{n} a_{k_2 j} x_j,$$

and taking the modulus, we get the bound

$$|a_{k_2 k_2} x_{k_2}| \leq R_{k_2}(A) |x_{k_1}|. \tag{2.9}$$

Similarly, using $(Ax)_{k_1} = 0$ gives the bound

$$|a_{k_1 k_1} x_{k_1}| \leq R_{k_1}(A) |x_{k_2}|, \tag{2.10}$$

and multiplying (2.9) and (2.10), we see that condition (2.7) is violated. $\square$

We are now in a position to prove Theorem 2.7: if λ is an eigenvalue of A, then the matrix $A - \lambda I$ is not invertible, and thus condition (2.7) is violated for this matrix. The deleted sums of the matrix $A - \lambda I$ are exactly the same as the deleted sums of the matrix A, so we get the inequality

$$|\lambda - a_{ii}||\lambda - a_{jj}| \leq R_i(A) R_j(A)$$

for some pair i, j. $\square$

How do Brauer's ovals compare to Gershgorin's discs? Writing

$$B_{ij}(A) := \{z \in \mathbb{C} : \quad |z - a_{ii}||z - a_{jj}| \leq R_i(A) R_j(A)\},$$

we see that Brauer's theorem is the inclusion

$$\sigma(A) \subset \bigcup_{i,j=1}^{n} B_{ij}(A). \tag{2.11}$$

Which of the two inclusions (2.11) and (2.6) is better? It should be easy to convince yourself that $B_{ij}(A) \subset G_i(A) \cup G_j(A)$ (think of the complements). So Brauer definitely wins (note that the so-called *Brauer oval* B_{ij} is used only for $i \neq j$, since for $i = j$ it would just amount to the disc B_i). Given that Gershgorin's discs are derived by examining the rows of the matrix A "one at a time" and Brauer's ovals "two at a time," one might think that it should be straightforward to keep going and move to products of three or more distances and so on. Unfortunately, the situation is a bit more complicated; such a generalization was indeed proved by Brualdi in [6], and it is formulated in terms of the loops of the graph $G(A)$; the interested reader may consult [6] and [35] for details.

2.3 Positive matrices and Perron's theorem

First some notation: if A is a matrix, we denote by $|A|$ the matrix made up of the moduli of the entries of A, meaning $|A|(i,j) := |A(i,j)|$ for all i, j. This includes vectors: if $y \in \mathbb{C}^n$, then $|y|_i = |y_i|$.

2 Linear algebra and search engines

Definition 2.9. *A matrix A with real entries is said to be*

nonnegative (written $A \geq 0$) if all its entries are nonnegative;
positive (written $A > 0$) if all its entries are positive.

Note that our definition is not restricted to square matrices, which means it includes vectors as well. A slight word of caution is in order here: unlike what happens for real numbers, the conditions $A \leq B$ and $A \neq B$ do not imply $A < B$; however, strict inequalities may be produced from nonstrict ones by multiplying by a positive matrix:

Lemma 2.10. *If A is a positive $n \times n$ matrix and x, y are two vectors satisfying $x \leq y$, $x \neq y$, then $Ax < Ay$.*

Proof. The ith component of the vector $Ay - Ax$ is given by

$$(Ay - Ax)_i = \sum_{j=1}^{n} A(i,j)(y_j - x_j).$$

This is clearly nonnegative, and it may vanish only if $y = x$. □

The notion of nonnegative matrix immediately gives us a partial order on $\mathcal{M}_n(\mathbb{R})$: we shall write $A \geq B$ when $A - B \geq 0$. Just as for real numbers, inequalities may be multiplied:

Lemma 2.11. *Let A, B, C, D be nonnegative matrices such that $A \leq B$ and $C \leq D$. Then $AC \leq BD$.*

As a consequence, if $0 \leq A \leq B$, then for every $k \in \mathbb{N}$ we have $A^k \leq B^k$. Our first result now tells us what matrix inequalities say about the spectrum:

Theorem 2.12. *Assume that $0 \leq A \leq B$. Then $\rho(A) \leq \rho(B)$.*

Proof. For all $k \in \mathbb{N}$ we know that $A^k \leq B^k$. Therefore, the Frobenius norms (see Appendix C) satisfy $|A^k|_F \leq |B^k|_F$. The conclusion now follows from the spectral radius formula (C.3). □

Our next result relates the spectral radius to row sums:

Theorem 2.13. *For a matrix $A \in \mathcal{M}_n(\mathbb{R})$, let $r_i(A)$ denote the ith row sum, and let $r(A), R(A)$ be the minimum and maximum of all row sums:*

$$r_i(A) := \sum_{j} a_{ij}, \quad r(A) := \min_{i} r_i(A), \quad R(A) := \max_{i} r_i(A).$$

Then we have $\rho(A) \leq R(A)$, and if $A \geq 0$, we also have $\rho(A) \geq r(A)$.

Note that as a corollary, this gives us the first half of Theorem 2.1. Indeed, for a Markov matrix all row sums are 1, and therefore, the spectral radius has to be 1.

Proof. The upper bound is immediate from (C.2) and Theorem C.2, so let us prove the lower bound.

If $r(A) = 0$, the lower bound is trivially valid, so let us assume that $r(A) > 0$. If we define a matrix B by $b_{ij} := \frac{r(A)}{r_i(A)} a_{ij}$, then clearly we have $0 \le B \le A$, so by Theorem 2.12 we have $\rho(B) \le \rho(A)$. Let $\mathbb{1} \in \mathbb{R}^n$ denote the vector all of whose entries are 1; you can check that $\mathbb{1}$ is an eigenvector of the matrix B associated with the eigenvalue $r(A)$. This tells us that $\rho(B) \ge r(A)$. On the other hand, considering the max-row-sum of B (see C.2), we have

$$\rho(B) \le |B|_\infty = r(A),$$

whence $\rho(B) = r(A)$. $\square$

An interesting consequence says that on multiplication by A, a vector cannot be expanded by more than a factor $\rho(A)$:

Theorem 2.14. *Let A be a positive $n \times n$ matrix, and x a nonnegative vector such that $Ax \ge \rho(A)x$. Then $Ax = \rho(A)x$.*

Proof. First note that from Theorem 2.13, we have $\rho(A) > 0$; let us assume that $Ax \ne \rho(A)x$, and set $z := Ax$. From Lemma 2.10 we have $Az > \rho(A)z > 0$, or in components,

$$\sum_{j=1}^{n} a_{ij} z_j > \rho(A) z_i > 0.$$

Let Z be the diagonal matrix made up of the components of z (note that $|z| > 0$, so Z is invertible). Then with the notation of Theorem 2.13, the row sums of the matrix $Z^{-1}AZ$ are given by

$$r_i(Z^{-1}AZ) = \frac{1}{z_i} \sum_j a_{ij} z_j,$$

and are therefore all strictly greater than $\rho(A)$. From Theorem 2.13 we get $\rho(A) > \rho(A)$, a contradiction. $\square$

We now have all the tools to proceed with the celebrated theorem of Perron:

Theorem 2.15 (Perron's theorem, 1907 [29]). *Let A be a positive $n \times n$ matrix. Then*

1. *$\rho(A) > 0$;*
2. *$\rho(A)$ is an eigenvalue;*
3. *there exists a positive eigenvector associated with the eigenvalue $\rho(A)$;*
4. *the eigenvalue $\rho(A)$ is dominant;*
5. *the eigenvalue $\rho(A)$ is geometrically simple, which means that the corresponding eigenspace is one-dimensional.*

Proof. As already noted, the lower bound $\rho(A) > 0$ is an immediate consequence of Theorem 2.13, since for a positive matrix we have $r(A) > 0$. To establish the next two points, let λ be an eigenvalue of maximum modulus, and x an associated eigenvector:

$$|\lambda| = \rho(A), \; Ax = \lambda x, \; x \neq 0.$$

Since A is positive, we have $A|x| \geq |Ax|$; on the other hand, we have $|Ax| = \rho(A)|x|$, and thus (combining the two) $A|x| \geq \rho(A)|x|$. Lemma 2.14 now tells us that $A|x| = \rho(A)|x|$, which shows that x is a positive vector, and is indeed an eigenvector of the matrix A associated with the eigenvalue $\rho(A)$. Thus points 2 and 3 are proved. In fact, our conclusion is slightly stronger: we have shown that every eigenvector x associated with an eigenvalue of maximum modulus satisfies $|x| > 0$ and $A|x| = \rho(A)|x|$; in other words, $|x|$ is an eigenvector associated with the eigenvalue $\rho(A)$ (recall that it is part of the definition of an eigenvector that it should be nonzero). This may even be pushed a little bit further: combining the relations $Ax = \lambda x$ and $A|x| = \rho(A)|x|$, we obtain $|Ax| = A|x|$, or in components,

$$\left| \sum_j a_{ij} x_j \right| = \sum_j a_{ij} |x_j| = \rho(A)|x_i|.$$

This (we have already encountered this type of collinearity argument in the proof of Theorem 2.3) tells us that the complex numbers $(a_{ij} x_j)_{1 \leq j \leq n}$ must be collinear, which means that (since the entries of A are real) the numbers $(x_j)_{1 \leq j \leq n}$ must be collinear:

$$x_j = e^{i\theta} |x_j|$$

for some $\theta \in \mathbb{R}$, or in vector form, $x = e^{i\theta}|x|$. Returning to $A|x| = \rho(A)|x|$, we obtain

$$Ax = e^{i\theta} A|x| = e^{i\theta} \rho(A)|x| = \rho(A)x,$$

and therefore, we must have $\lambda = \rho(A)$, which means that the eigenvalue $\rho(A)$ is dominant. Finally, let us show that the geometric multiplicity of $\rho(A)$ is 1. Choosing two eigenvectors x and y, we now know that $x = e^{i\theta}|x|$, $y = e^{i\mu}|x|$ for some real numbers θ, μ, and that all components of x and y are nonzero. Let α be defined by

$$\alpha := \min_i \left| \frac{x_i}{y_i} \right|,$$

and set $z := x - \alpha y$. Then you may check that $Az = \rho(A)z$. If z were not zero, this would mean that z qualified as an eigenvector associated with $\rho(A)$, and as such, all its components would have to be nonzero. But this is not the case, since from the very definition of α it follows that at least one component of the vector z vanishes. Therefore, $z = 0$, which means that $|x|$ and $|y|$ are collinear. Therefore, x and y are collinear. $\square$

It turns out that the last point in the conclusion of Perron's theorem may be sharpened (recall from Appendix C that the geometric multiplicity is always less than or equal to the algebraic multiplicity):

Theorem 2.16. *Let A be a positive $n \times n$ matrix. Then the eigenvalue $\rho(A)$ is simple.*

Proof. Let us denote by p_A the characteristic polynomial of A:

$$p_A(X) = \det(XI_n - A) = \det(Xe_1 - C_1, \ldots, Xe_n - C_n).$$

Here e_k denotes the kth basis vector, and C_k denotes the kth column vector of the matrix A. We know that $p_A(\rho(A)) = 0$, and we need to prove that $p'_A(\rho(A)) \neq 0$. By multilinearity the derivative p'_A may be computed as follows:

$$p'_A(X) = \sum_{k=1}^{n} \det(Xe_1 - C_1, \ldots, e_k, \ldots, Xe_n - C_n) \qquad (2.12)$$

$$= \sum_{k=1}^{n} p_{A_k}(X), \qquad (2.13)$$

where A_k is the $(n-1) \times (n-1)$ positive matrix obtained from A by erasing the kth column and kth row. We are now going to show that $p_{A_k}(\rho(A))$ is positive for all k. Let B_k be the $n \times n$ matrix obtained from A_k by addition of a zero column on the right and a zero row on the bottom:

$$B_k := \begin{pmatrix} A_k & 0 \\ 0 & 0 \end{pmatrix}. \qquad (2.14)$$

Then $p_{B_k}(X) = Xp_{A_k}(X)$, and thus the spectral radii are equal: $\rho(B_k) = \rho(A_k)$. For every eigenvector $x \in \mathbb{R}^{n-1}$ of A_k and real number $y \in \mathbb{R}$, the vector $z := (x, y) \in \mathbb{R}^n$ is an eigenvector of B_k, associated with the same eigenvalue. Perron's theorem applied to A_k gives us a positive eigenvector associated with the eigenvalue $\rho(A_k)$, which then gives us a positive eigenvector z of B_k:

$$B_k z = \rho(A_k)z, \quad z > 0.$$

Since $0 \leq B_k \leq A$, from Theorem 2.12 we have $\rho(B_k) \leq \rho(A)$. We are going to show that $\rho(B_k) < \rho(A)$. Assuming equality, we have $B_k z = \rho(A)z$, which (recall that both A and z are positive) implies $Az \geq \rho(A)z$. From Theorem 2.14 we obtain $Az = \rho(A)z$, and thus $Az = B_k z$. This is impossible, since $Az > 0$, whereas the vector $B_k z$ has at least one zero component. This shows that $\rho(A_k) < \rho(A)$. Since $\rho(A_k)$ is the largest real eigenvalue of A_k, we conclude that $p_{A_k}(\rho(A))$ has the sign of p_{A_k} near infinity, $p_{A_k}(\rho(A)) > 0$, which shows that $p'_A(\rho(A)) > 0$. $\square$

2.4 Irreducibility and the theorem of Frobenius

If one wishes to generalize Perron's theorem to nonnegative matrices, it is easy to see with a few examples what can go wrong. For instance, the matrix

$$\begin{pmatrix} 0 & 1 \\ 0 & 0 \end{pmatrix} \tag{2.15}$$

has only the eigenvalue $\lambda = 0$. This eigenvalue has algebraic multiplicity 2, and there is no positive associated eigenvector. In the absence of additional conditions, the following is all that can be salvaged for general nonnegative matrices:

Theorem 2.17. *Let A be a nonnegative matrix. Then*

1. *$\rho(A)$ is an eigenvalue;*
2. *there exists a nonnegative eigenvector associated with the eigenvalue $\rho(A)$.*

Proof. The proof relies on approximating A by positive matrices and a compactness argument. More precisely, for every $k \geq 1$ let A_k be defined by $A_k(i,j) := A(i,j) + \frac{1}{k}$. Then A_k is a positive matrix, and thus Perron's theorem gives us a positive vector x^k satisfying $A_k x^k = \rho(A_k) x^k$. We may normalize it and set $y^k := \frac{x^k}{||x^k||}$ (using whatever vector norm $|| \cdot ||$ we please). Each y^k is an eigenvector of A_k associated to the eigenvalue $\rho(A_k)$, and since the sequence $(y^k)_{k \in \mathbb{N}}$ is bounded, we may extract a subsequence $(y^{k_l})_{l \in \mathbb{N}}$ that converges (in the vector norm we chose to use) to some nonnegative unit vector y. Since $(A_k)_{k \in \mathbb{N}}$ converges to A and $(y^{k_l})_{l \in \mathbb{N}}$ converges to y, it follows that $(A_{k_l} y^{k_l})_{l \in \mathbb{N}}$ converges to Ay; that is, $(\rho(A_{k_l}) y^{k_l})_{l \in \mathbb{N}}$ converges to Ay:

$$(\rho(A_{k_l}) y^{k_l})_{l \in \mathbb{N}} \to Ay. \tag{2.16}$$

(This is an easy exercise, which you should try. Hint: in the relevant subordinate matrix norm, A_k converges to A.) The vector y is not zero (it is a unit vector), so if we choose i such that $y_i > 0$, then for l large enough we have $y_i^{k_l} > 0$, and therefore, from (2.16) we see that $\rho(A_{k_l})$ must have a limit ρ. Taking into account the convergence of y^{k_l}, from (2.16) we obtain $Ay = \rho y$, and thus ρ being an eigenvalue of A, we must have $\rho \leq \rho(A)$. On the other hand, since $0 \leq A \leq A_k$, from Theorem 2.12 we have $\rho(A_k) \geq \rho(A)$, and thus $\rho \geq \rho(A)$. Finally, we have shown that $\rho = \rho(A)$ and that $\rho(A)$ is an eigenvalue of A. $\square$

How can we weaken the assumption of positivity and still retain some of the other conclusions of Perron's theorem? One might think that the matrix given by (2.15) has in a sense too many zeroes for any kind of generalization of Perron's theorem to apply; in fact, as we shall see, it is not only the number of zero entries in the matrix that plays a role here, but their location. Let us, for instance, consider the two matrices

$$A := \begin{pmatrix} 1 & 0 \\ 1 & 1 \end{pmatrix}, \quad B := \begin{pmatrix} 1 & 1 \\ 1 & 0 \end{pmatrix}.$$

You will easily check that A has only the eigenvalue 1, which has multiplicity 2, and that there is no positive eigenvector (more precisely, you will see that every eigenvector x satisfies $x_1 = 0$). On the other hand, the spectral radius of B is $\frac{1+\sqrt{5}}{2}$, which is a simple and dominant eigenvalue, the corresponding eigenspace being generated by the positive vector $(\frac{1+\sqrt{5}}{2}, 1)$. A good way to visualize what is going on here is to look at the corresponding graphs. Given a nonnegative $n \times n$ matrix A, its directed graph $G(A)$ is defined by connecting i to j whenever $A(i,j) \neq 0$. In other words, $G(A)$ has vertex set $\{1, \ldots, n\}$ and edge set all pairs (i,j) such that $A(i,j) \neq 0$. Here we obtain the following picture:

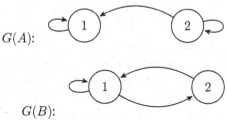

The fact that every eigenvector of A must satisfy $x_1 = 0$ is due to the equality $A(1,1) = 0$, which here visually corresponds to the fact that the point 1 loops on itself and does not lead to the point 2; in graph-theoretic terms this means that $G(A)$ is not strongly connected; let us recall what this means:

Definition 2.18. *A directed graph $G = (V, E)$ is said to be strongly connected (or path connected) if for each pair of vertices $V_i, V_j \in V$ there is a finite sequence of edges leading from V_i to V_j.*

Note that in the example above the directed graph $G(B)$ is strongly connected, whereas $G(A)$ is not.

At this point we are led to conjecture that if we restrict ourselves to the class of nonnegative matrices A for which $G(A)$ is strongly connected, then some generalization of Perron's theorem might be possible; this is exactly what was proved by Frobenius in 1912. However, in order to state Frobenius's theorem as a result strictly about matrices (i.e., without having to refer to the graph $G(A)$), we need a characterization of such matrices; a slight detour through paths and graphs is required before we can obtain this characterization, and we postpone the statement and proof of Frobenius's theorem to the next section:

Definition 2.19. *For a pair of vertices $V_i, V_j \in V$ of a directed graph $G = (E, V)$, we say that there exists a 1-path from V_i to V_j if $(V_i, V_j) \in E$, and for $m \geq 2$, we say that there exists an m-path from V_i to V_j if one can find $m - 1$ indices $k_1, \ldots, k_{m-1}$ such that $(V_i, V_{k_1}), (V_{k_1}, V_{k_2}), \ldots, (V_{k_{m-1}}, V_j)$ are all edges of G.*

Note that in the case of the directed graph $G(A)$ of a matrix A, this condition means that the product $A(i, k_1)A(k_1, k_2) \cdots A(k_{m-1}, j)$ is not zero; here, however, is a more useful criterion:

Lemma 2.20. *For a pair $i, j \in \{1, \ldots, n\}$ and integer $m \geq 1$, there exists an m-path from i to j in the directed graph $G(A)$ if and only if $A^m(i, j) \neq 0$.*

Proof. This being obviously true for $m = 1$, let us keep going by induction; assuming that it is true for m, we may write

$$A^{m+1}(i,j) = \sum_{k=1}^{n} A^m(i,k)A(k,j).$$

This shows that $A^{m+1}(i,j) \neq 0$ if and only if for some k we have $A^m(i,k)A(k,j) \neq 0$, i.e., if and only if there exists an m-path from i to k and a 1-path from k to j. This is exactly equivalent to saying that there exists an $(m+1)$-path from i to j, which completes the induction step. $\square$

Strong connectivity of the directed graph $G(A)$ means that for every pair i, j there is an m-path from i to j for some $m \geq 1$; of course, if m is larger than $n - 1$, then such a path has to visit some vertex at least twice and may therefore be shortened. This will be quite useful in what follows:

Remark 2.21. A directed graph $G(A)$ is strongly connected if and only if for every pair of vertices $i, j \in \{1, \ldots, n\}$ there is an m-path leading from i to j, for some $m \leq n - 1$.

Our first characterization of matrices A for which $G(A)$ is not strongly connected is the following:

Theorem 2.22. *Let A be a nonnegative matrix, and $G(A)$ its directed graph. Then $G(A)$ is not strictly connected if and only if for some permutation matrix P the matrix $P^t AP$ is block upper triangular, which means that it is of the form*

$$P^t AP = \begin{pmatrix} X & Y \\ 0 & Z \end{pmatrix}, \tag{2.17}$$

where X and Z are square matrices.

Proof. It follows from (C.7) that permutation similarity does not change the graph; more precisely, for every permutation $\sigma \in \mathfrak{S}_n$, the directed graphs $G(A)$ and $G(P_\sigma^t AP_\sigma)$ consist of the same vertices and edges (the only difference being in the numbering of the vertices). It follows that $G(A)$ is strongly connected if and only if $G(P_\sigma^t AP_\sigma)$ is. Let us assume that (2.17) holds for some permutation matrix P, and let s be the size of the zero block in the lower left corner; this means that the first s vertices of the graph $G(P_\sigma^t AP_\sigma)$ are not accessible from the last $n - s$ vertices, which means that $G(P_\sigma^t AP_\sigma)$ is not strongly connected, so neither is $G(A)$. Conversely, let us assume that $G(A)$ is not strongly connected. This means that some vertex V_{i_2} of $G(A)$ is not accessible from some other vertex V_{i_1}. (It must be true for some pair i_1, i_2 with $i_1 \neq i_2$; can you see why?) In particular, we must have $A(i_1, i_2) = 0$. It remains to bring this 0 entry to the lower left corner by renumbering the vertices. More precisely, let $\sigma \in \mathfrak{S}_n$ be the permutation defined by $\sigma(i_1) = n, \sigma(i_2) = 1$ and leaving all other indices fixed. Then the matrix $B := P_\sigma^t A P_\sigma$ is block upper triangular (note that if there are other unreachable points, we might keep going to increase the size of the zero block, but the result is already proved with a block of size 1). □

Obviously, such matrices must be given a name:

Definition 2.23. *An $n \times n$ matrix A is said to be reducible if (2.17) is true for some $n \times n$ permutation matrix P; otherwise, A is said to be irreducible.*

Now that we have understood the relevance of this property, we would like to have simple criteria for (ir)reducibility; one first condition is a direct consequence of Lemma 2.20:

Theorem 2.24. *For a nonnegative matrix A the following three conditions are equivalent:*

1. *A is irreducible;*
2. *for every pair of indices i, j there exists an integer k, $1 \leq k \leq n-1$, such that $A^k(i, j) > 0$;*
3.
$$(I + A)^{n-1} > 0.$$

Proof. The equivalence of 1 and 2 follows at once from Lemma 2.20 and Remark 2.21. To show that 1 implies 3 we use the binomial formula:

$$(I + A)^{n-1} = \sum_{k=0}^{n-1} \binom{n-1}{k} A^k.$$

It follows from 2 that the matrix on the right-hand side in this expression is positive, which establishes 3. Finally, to show that 3 implies 1, let us argue by contradiction: if A were reducible from (2.17), we would have

$$(I + A)^k = P \begin{pmatrix} (I + X)^k & Y_k \\ 0 & (I + Z)^k \end{pmatrix} P^t$$

for every $k \geq 1$ (here Y_k is some block that we do not need to compute exactly), which means that no power of the matrix $I + A$ can be positive. □

The conditions of Theorem 2.24 are not met when some vertices cannot be joined to others, which must manifest itself in the presence of zeroes in the matrix A; the precise formulation is the following:

Theorem 2.25. *A nonnegative matrix A is reducible if and only if the set of indices $\{1, \ldots, n\}$ is the disjoint union of two nonempty sets I and J such that for every $i \in I$, $j \in J$ we have $A(i, j) = 0$.*

Proof. If A is reducible, take σ to be the permutation associated with the matrix P in (2.17), and let s be the size of the square block of zeroes. Then we have

$$(P_\sigma^t A P_\sigma)(i,j) = 0 \text{ for any } i \in \{n - s + 1, \ldots, n\}, j \in \{1, \ldots, s\}. \qquad (2.18)$$

This shows that A is reducible if and only if (2.18) is true for some $s \geq 1$ and some $\sigma \in \mathfrak{S}_n$. Defining the sets I and J by

$$I := \{\sigma^{-1}(n - s + 1), \ldots, \sigma^{-1}(n)\}, \ J := \{\sigma^{-1}(1), \ldots, \sigma^{-1}(s)\}$$

and using (C.7), we obtain the desired characterization. $\square$

Let us summarize the various equivalent formulations of reducibility:

Theorem 2.26. *Among nonnegative matrices, reducible matrices are characterized by the following five conditions, which are all equivalent:*

1. *$G(A)$ is not strongly connected;*
2. *there exists a pair i, j such that $A^k(i,j) = 0$ for every $k \geq 1$;*
3. *the set $\{1, \ldots, n\}$ may be partitioned into two subsets I, J such that $A(i,j) = 0$ for all $i \in I$, $j \in J$;*
4. *(2.17) holds for some permutation matrix P;*
5. *the matrix $(I + A)^{n-1}$ has some zero entries.*

Some of these conditions were met in negative form above, and they characterize irreducible matrices.

We now have all the material needed for Frobenius's generalization of Perron's theorem (the union of both theorems is sometimes referred to as *the Perron–Frobenius theorem*):

Theorem 2.27 (Frobenius's theorem, 1912 [12]). *Let A be a nonnegative irreducible matrix. Then all conclusions of Perron's theorem are valid, except (perhaps) for point 4; moreover, the eigenvalue $\rho(A)$ is simple.*

Proof. To establish the first point, from Theorem 2.13 it suffices to show that all row sums of A are positive. Arguing by contradiction, assume that some row sum $r_k(A)$ of A vanishes (which since A is nonnegative means that all entries in the kth row of A are zero); using a suitable permutation, this row may be made to be the last one in a square matrix similar to A. More precisely, for every permutation $\sigma \in \mathfrak{S}_n$ such that $\sigma(n) = k$, the matrix $P_\sigma^t A P_\sigma$ has an nth row made up entirely of zeroes, which means that A is reducible. Point 2 is known from Theorem 2.17, and in fact, we are going to show that the nonnegative nonzero vector y provided by Theorem 2.17 is positive (thus establishing point 3). From $Ay = \rho(A)y$ and the binomial formula we obtain the relation

$$(I + A)^{n-1}y = (1 + \rho(A))^{n-1}y. \qquad (2.19)$$

We know from Theorem 2.24 that the matrix $(I + A)^{n-1}$ is positive; therefore (since $y \neq 0$), the vector $(I + A)^{n-1}y$ is positive, and from (2.19) we see that $y > 0$. Finally, let us show that the eigenvalue $\rho(A)$ is algebraically (and therefore geometrically) simple. The spectrum of $(I + A)^{n-1}$ is given by

$$\sigma((I + A)^{n-1}) = \{(1 + \lambda)^{n-1}, \lambda \in \sigma(A)\}.$$

We already know that $\rho(A)$ is an eigenvalue of A; therefore, $(1 + \rho(A))^{n-1}$ is an eigenvalue of $(I + A)^{n-1}$, which immediately gives the lower bound $\rho((I + A)^{n-1}) \geq (1 + \rho(A))^{n-1}$. On the other hand, we have the upper bound

$$\rho((I + A)^{n-1}) = \max\{|(1 + \lambda)^{n_1}|, \lambda \in \sigma(A)\} \leq (1 + \rho(A))^{n-1}$$

(here again we use the binomial formula to expand $(1 + \lambda)^{n-1}$), so finally we have equality: $\rho((I + A)^{n-1}) = (1 + \rho(A))^{n-1}$. Since the matrix $(I + A)^{n-1}$ is positive, from Theorem 2.16 its spectral radius $(1 + \rho(A))^{n-1}$ is an algebraically simple eigenvalue. From Theorem C.3, there exists a unitary matrix U such that U^*AU is upper triangular. Then the matrix $U^*(I + A)^{n-1}U$ also is upper triangular, and for every diagonal element λ in U^*AU the corresponding diagonal element in $U^*(I + A)^{n-1}U$ is $(1 + \lambda)^{n-1}$. The number $(1 + \rho(A))^{n-1}$ can occur only once on the diagonal of $U^*(I + A)^{n-1}U$, since it is an algebraically simple eigenvalue; it follows that $\rho(A)$ occurs only once on the diagonal of U^*AU, which means precisely that it is an algebraically simple eigenvalue of A. □

2.5 Primitivity and the power method

So far we have seen that irreducible nonnegative matrices provide a natural enlargement of the class of positive matrices for a generalization of Perron's theorem, and the only property lost along the way is the dominant character of the eigenvalue $\rho(A)$. This leads to the question of the characterization of square matrices for which the spectral radius is a dominant eigenvalue. Let us begin by giving them a name:

Definition 2.28. *A nonnegative irreducible square matrix A is said to be primitive if $\rho(A)$ is a dominant eigenvalue of A.*

We shall see that primitive matrices are exactly matrices for which (after normalization by the spectral radius) successive powers have a nontrivial limit; this is the essence of the so-called *power method*, which uses successive powers of the matrix to compute an approximation of the positive eigenvector associated with $\rho(A)$:

Theorem 2.29 (The power method for primitive matrices). *Let A be a nonzero primitive matrix A, and let $x > 0$ be a column vector such that $Ax = \rho(A)x$, and let $y > 0$ be a row vector satisfying $yA = \rho(A)y$ (such vectors exist by Theorem 2.27 applied to A and A^t). Then*

$$\lim_{k \to \infty} [\frac{A}{\rho(A)}]^k = \frac{xy}{yx}.$$

Proof. From Theorem 2.27 we know that $\rho(A)$ is a simple eigenvalue of A. Thus (since it is also dominant) the result follows immediately from Theorems C.7 and C.8. □

Remark: it may be shown that the convergence stated in the theorem is in fact a characterization of primitive matrices within the set of nonnegative irreducible matrices; see, for instance, [25] p. 674.

Of paramount importance is the special case of primitive Markov matrices:

Theorem 2.30. *Let P be a primitive Markov matrix, and let π be its unique stationary distribution. Then for every probability vector p we have*

$$\lim_{k \to \infty} pP^k = \pi.$$

Proof. We know from Theorem 2.1 that $\rho(P) = 1$, and for this eigenvalue we have (for free) the right eigenvector $x := (1, \ldots, 1)^t$, the column vector made up of 1's only. Frobenius's theorem applied to the matrix P^t gives us a positive right eigenvector for P^t, or in other words, a positive left eigenvector y for P. Since the eigenvalue 1 is simple, the associated eigenspace is one-dimensional. In the language of Markov chains this means (dividing y by the sum of its components) that there exists a unique stationary distribution π. Now the convergence in Theorem 2.29 tells us that as $k \to \infty$, we have $P^k \to x\pi$ (note that a vector p is a probability vector p if and only if $px = 1$; therefore, $\pi x = 1$). For every probability vector p we have $px\pi = \pi$, and thus as $k \to \infty$ we indeed have $pP^k \to \pi$. □

Another important consequence of Theorem 2.29 is the celebrated criterion for primitivity of Frobenius:

Theorem 2.31 (Frobenius's test for primitivity). *A nonnegative matrix A is primitive if and only if some power of A is positive: $A^k > 0$ for some $k \in \mathbb{N}$.*

A simple example that fails the test is the matrix given by

$$\begin{pmatrix} 0 & 1 \\ 1 & 0 \end{pmatrix}. \tag{2.20}$$

It is clearly irreducible, but not primitive (compute its successive powers). Note that A has two eigenvalues of maximum modulus: $\sigma(A) = \{-1, 1\}$.

Proof. If A is primitive, the convergence in Theorem 2.29 tells us (since the limit $\frac{xy}{yx}$ is a positive matrix) that for k large enough, all elements of A^k are positive. Conversely, assume that $A^k > 0$ and that the matrix A has r eigenvalues of maximum modulus:

$$|\lambda_1| = \cdots = |\lambda_r| = \rho(A).$$

Then the numbers $\lambda_1^k, \ldots, \lambda_r^k$ are eigenvalues of maximum modulus of the matrix A^k. From Perron's theorem there can be only one such eigenvalue. Therefore, we obtain

$$\lambda_1^k = \cdots = \lambda_r^k = \rho(A)^k.$$

In the triangular matrix $U^* A U$ provided by Theorem C.3, each of the eigenvalues $\lambda_1, \ldots, \lambda_r$ appears on the diagonal, the number of occurrences being its algebraic multiplicity. Passing to the triangularized form of A^k, these give diagonal entries in the matrix $U^* A^k U$ that are all equal to $\rho(A)^k$, but since (from Theorem 2.16) this eigenvalue of A^k is simple, there can be only one such occurrence; it follows that $r = 1$. $\square$

Now comes one of those rare moments when seemingly unrelated things come together: the translation of this condition in the language of Markov chains (in the case that A is a Markov matrix) is ergodicity of the chain, as shown by Theorem 1.48. This understood, it is now clear that Theorem 2.30 is nothing other than the linear-algebraic formulation of the convergence result for ergodic chains, Theorem 1.39. Here is yet another way to say this: we have proved the same result twice, once (in the proof of Theorem 1.39) by probabilistic methods, once (in the proof of Theorem 2.30) by linear-algebraic methods.

As befits a section devoted to graphs and matrices, let us close with an oriented graph that summarizes the interdependence among various positivity and spectral properties of nonnegative square matrices; the contents of any box may be (and is in some texts) taken as a definition of the label on top of it.

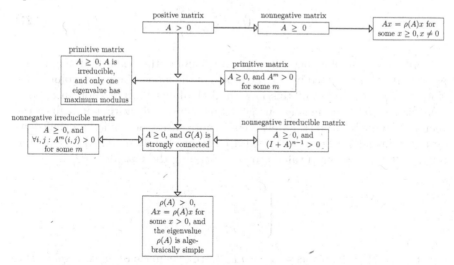

2.6 The PageRank algorithm

2.6.1 Formulation as an eigenvalue problem

The ever increasing size of the World Wide Web makes it necessary for search engines to rely on some form of link-based hierarchy of the pages. Roughly speaking, the issue is to assign to each page some quantitative measure of its "importance," which we shall call its rank, so that the results of a query may be provided by order of decreasing rank. The PageRank method, first published in [28], is one the first—and probably the most famous—models for defining and computing page ranks (for a very clear presentation of the history of link-based ranking as well as an excellent introduction to the PageRank algorithm, we refer the reader to [3] and [17], on which most of this section is based; for further detail see also the monograph [18]). Consider the internet as an extremely large oriented graph in which each node is a page, and an edge connects page i to page j if and only if a link to page j is present on page i, which we will indicate by writing $i \to j$. The number of such pages directly accessible from page i will be denoted by d_i. Now the goal is to define a vector r such that for each page i, the quantity $r(i)$ is a "good measure" of its "importance." At the basis of the implicit definition of r are two ideas: first, a page i is important if it is accessible from many pages; second, the contribution of each j such that $j \to i$ should be weighted: if a page j has many outgoing links (meaning if d_j is large), then it should not contribute too much to the numerical value of $r(i)$. Combining these two ideas leads to requiring the following:

$$\forall i : \quad r(i) = \sum_{j \to i} \frac{r(j)}{d_j}. \tag{2.21}$$

In words: the rank of a page is the weighted sum of the ranks of all pages that point to it, each weighted by the reciprocal of the number of its outgoing links. Note that for the moment this is wishful thinking: (2.21) is not an explicit definition for $r(i)$, so ideally our goal would be to show that it defines the vector r uniquely. Unfortunately, this is not quite true, and the coefficients on the right-hand side of (2.21) will need to be slightly modified. Let us first rewrite (2.21) as an eigenvalue problem: defining the *link matrix* P by

$$P(i,j) = \begin{cases} 0 & \forall j \quad \text{if} \quad d_i = 0; \\ \frac{1}{d_i} & \text{if } d_i \neq 0 \text{ and } i \to j; \\ 0 & \text{otherwise,} \end{cases}$$

equation (2.21) is equivalent to $r = rP$. In the language of linear algebra this means that P should have 1 as an eigenvalue, and r should have to be a left eigenvector of the matrix P associated with that eigenvalue. In the language of Markov chains, r would have to be a stationary distribution for a Markov chain that evolves as follows: when present on a page i, a surfer draws his next position uniformly from the d_i pages present on i; of course this makes no sense if $d_i = 0$, a problem we shall need to fix.

As mentioned above, this is just wishful thinking for the moment: in order to be able to use what we know on the existence of stationary distributions for Markov chains, we need to modify P so as to obtain a primitive Markov matrix. The by now classical presentation of the PageRank method does this in two steps.

Step 1: the "Markov" fix. The matrix P fails to be a Markov matrix because some of its row sums are not 1: the ith row sum is d_i, which is 0 if no outgoing link is present on page i. Such pages are called *dangling nodes*. The simplest way to modify the rows of P corresponding to dangling nodes is simply to set all entries equal to $\frac{1}{n}$, where n is the total number of pages. In random walk language this means that when present on a page with no outgoing link, our surfer draws his next position uniformly from all the existing pages. Mathematically, if we define the column vector a by $a_i = 0$ if i is dangling and $a_i = 0$ otherwise, our new matrix is now

$$P' := P + \frac{1}{n}ae^t, \tag{2.22}$$

where e is the column vector with all entries equal to 1. The matrix P' is stochastic, but it is extremely sparse, meaning that it has a large number of zeroes, so it is rather unlikely to be primitive.

Step 2: the "primitivity" fix (although it should really be called the positivity fix). In order to get a positive Markov matrix we take a convex combination of P' and the so-called *teleportation matrix*, i.e., the matrix $T := \frac{1}{n}ee^t$ with all entries equal to $\frac{1}{n}$. In other words, we set

$$P'' := \alpha P' + (1 - \alpha)\frac{1}{n}ee^t, \tag{2.23}$$

where $\alpha \in (0, 1)$ is some chosen parameter. Then P'' is both stochastic and positive, and Perron's theorem gives us the existence and uniqueness of a probability vector r satisfying $r = rP''$; we shall call it the PageRank vector. Note that r is obtained from P'', so an important issue is how far it is from satisfying the original problem (2.21); this is determined by how far the matrix P'' is from P; combining the two steps, we see that P'' is a rank-one perturbation of P:

$$P'' = \alpha P + \frac{1}{n}[\alpha a + (1 - \alpha e)]e^t.$$

Roughly speaking, P'' is close to P if there are not too many dangling nodes (so step 1 did not perturb things too much) and α is close to 1 (so step 2 did not do so either). How can we compute a good approximation to r? Given that P'' is positive (therefore primitive), we may use the power method as in Theorem 2.30:

Theorem 2.32. *The matrix P'' defined from the link matrix P by (2.22) and (2.23) is a positive Markov matrix for every $\alpha \in (0, 1)$. Let r be the unique probability vector r satisfying $r = rP''$. For a probability vector p, the sequence $(p_k)_{k \in \mathbb{N}}$ defined by $p_0 = p$, $p_{k+1} = p_k P''$ converges to r.*

2.6.2 Formulation as a linear system

Theorem 2.32 does two things: on the theoretical side it tells us that the problem $rP'' = r$ is well posed, and on the practical side it gives us an algorithm (the power method) for computing a good approximation to its solution. An alternative approach is to go back to the matrix P' and formulate the problem as a linear system with a given right-hand side; let us rewrite the condition $rP'' = r$ as follows:

$$r\left[\alpha P' + (1 - \alpha)\frac{ee^t}{n}\right] = r \iff r(I - \alpha P') = \frac{1}{n}(1 - \alpha)ree^t.$$

Noting (as in the proof of Theorem 2.30) that (if r is to be a probability vector) $re = 1$, we get the condition

$$r(I - \alpha P') = \frac{1}{n}(1 - \alpha)e^t. \tag{2.24}$$

For a probability vector this condition is equivalent to $rP'' = r$, which we already know is a well-posed problem; here is a nice approach to well-posedness based on M matrices:

Proposition 2.33. *The matrix $I - \alpha P'$ is an M matrix.*

This is an immediate consequence of Corollary C.14, since P' is a Markov matrix. By Theorem C.9, this implies that it is invertible and has a nonneg-• ative inverse; thus r is uniquely defined and is a nonnegative vector.

Let us mention that one more important issue is that of robustness of our solutions with respect to the data, i.e., the link matrix: if the link matrix is not exactly known, how does this affect our solution? More mathematically put, for two different link matrices P_1 and P_2, can we bound (in some norm) the difference $r_1 - r_2$ between the two corresponding solutions in terms of $P_1 - P_2$? It turns out that the formulation (2.24) provides an easy answer to this question; namely, the condition number of the matrix $I - \alpha P'$ may be determined exactly. For details, see, for instance, [17], p. 350.

2.7 Problems

2.1. Let A be a nonzero nonnegative $n \times n$ matrix. Assume that A has a positive row eigenvector for some eigenvalue λ, and a nonnegative eigenvector for some eigenvalue μ. Show that necessarily $\lambda = \mu$.

2.2. Show that every nonnegative square matrix A is irreducible if and only if A^t is irreducible; show that A is primitive if and only if $I + A$ is primitive.

2.3. Let A be a nonnegative irreducible matrix with all diagonal elements positive: $a_{ii} > 0$ for all i. Show that A is primitive.

2.4. Show without doing any computation that every 2×2 nonnegative matrix has two real eigenvalues.

2.5. Within the set of monotone 2×2 matrices, give a characterization of M matrices in terms of their determinant.

2.6. The Leontief model is a linear model of a closed economy that may be roughly described as follows. The system consists of n sectors numbered from 1 to n, and each sector produces exactly one good (call it its *output*).

For simplicity assume that all quantities are measured in the same units, and let a_{ij} denote the amount of product i needed in the production of one unit of product j; the so-called *technology matrix* A is assumed to be nonnegative. With x_i denoting the quantity produced by sector i, the internal consumption of the system in sector i is then $\sum_{j=1}^n a_{ij} x_j$. Thus its net production is $x_i - \sum_{j=1}^n a_{ij} x_j$. If this is to meet a demand vector d exactly, we get the linear model

$$(I - A)x = d.$$

The model is economically roadworthy if this has a nonnegative solution for all $d \geq 0$; the matrix A is said to be *productive* if there exists a vector $x > 0$ such that $(I - A)x > 0$.

1. Show that if all row sums of a matrix A are strictly less than 1, then A is productive.
2. Show that every productive matrix A is convergent, i.e., $A^n \to 0$ as $n \to \infty$.
3. Since a productive matrix is convergent, Theorems C.5 and C.6 tell us that it is monotone; for a productive matrix can you show monotonicity directly, i.e., without appealing to these results?

2.7. If you take a close look at the proof of the last point in Theorem 2.27, you should be able to prove the following statements about the eigenvalues of a polynomial of a square matrix:

1. For every $n \times n$ matrix A and polynomial P (with arbitrary complex coefficients), the algebraic multiplicity $m_A(\lambda)$ of the eigenvalue $\lambda \in \sigma(A)$ is less than or equal to the algebraic multiplicity of the eigenvalue $P(\lambda) \in \sigma(P(A))$, i.e., $m_A(\lambda) \leq m_{P(A)}(P(\lambda))$; can you think of a condition on P and $\sigma(A)$ that guarantees that they are always equal?
2. If an $n \times n$ matrix A is such that $\rho(A) \in \sigma(A)$, then for every polynomial P with positive coefficients we have $\rho(P(A)) = P(\rho(A))$.

2.8. Let us revisit the Dmitriev–Dynkin theorem in dimension two. Show by direct computation of the eigenvalues that the union Λ_2 of all spectra of all 2×2 stochastic matrices is the interval $[-1, 1]$.

3

The Poisson process

Summary. This chapter serves as an interlude between the discrete-time theory we just reviewed and the continuous-time theory. The Poisson process being probably the best (for pedagogical purposes) prototype of continuous-time Markov process, it provides a good occasion to gently introduce the concepts specific to this framework (such as right constancy, jump times, jump chain, properties of increments) before hitting the reader with the general theory, which will be done in the next chapter. At the heart of our approach to time-continuous processes is time discretization; the proofs of Theorems 3.9 and 3.10 will provide a gentle introduction to this method, which will be used again in the next chapter (with unfortunately many technical complications in the general case, where increments are not assumed to be stationary).

3.1 Some heuristics on jumps and consequences

For a continuous-time process the idea of memory loss as time increases may be formalized as in the discrete-time case (see Definition 4.1 in the next chapter); however, some complication arises here due to the fact that t is continuous: if the system starts in state $x \in S$ at time 0, then most likely it will not stay there forever, and it will at some later time "jump" to some other state $y \neq x$. When may this happen? If the process is "reasonable," surely there should be a smallest time T called the *first jump time* for which this happens; an "unreasonable" process would be one for which trajectories constantly jump from one state to another. In order to avoid this sort of erratic behavior, we will require some form of regularity for the paths that will ensure $T > 0$.

If S is a given countable set, it does not make sense to say that a map $f : [0, \infty) \to S$ is continuous unless S is endowed with a topology (one always tends to think of discrete spaces as subsets of $\mathbb{Z}$, but here S could be a set of political parties, or of vegetables, for which there is no notion of convergence). This is why instead of talking about continuous trajectories we will formalize this idea of regularity of paths by requiring the following condition:

Definition 3.1. *Let E be any set (countable or not), and $I \subset \mathbb{R}$ some interval. A map $f : I \to E$ is said to be right constant (r.c. for short) if for every $x \in I$*

© Springer International Publishing AG, part of Springer Nature 2018
J.-F. Collet, *Discrete Stochastic Processes and Applications*,
Universitext, https://doi.org/10.1007/978-3-319-74018-8_3

there exists $\epsilon > 0$ such that $x + \epsilon \in I$ and f is constant on the interval $[x, x + \epsilon)$. A process $(X_t)_{t>0}$ with values in E is said to be right constant if its trajectories are all right constant maps.

In some texts these maps are called *right continuous*, although no explicit mention is made of the topology used on the range of f. For more details on the connection between right constancy and right continuity, see Section A.4.

We can now be a little more precise in the description of the jumps we mentioned before: the *first jump time* of X is defined as the first time for which the state of the system differs from the initial one:

$$T_1 := \inf\{t \geq 0 : X_t \neq X_0\}.$$

As usual, we use the convention that the infimum of an empty set is equal to $+\infty$. In other words, if the process stays in the constant state X_0 forever, the first jump time is infinite. States for which this happens surely will be called *absorbing*:

Definition 3.2. *The state $x \in S$ is said to be absorbing if*

$$P(T_1 = \infty | X_0 = x) = 1.$$

The subsequent jump times are then defined recursively:

Definition 3.3. *If $X = (X_t)_{t \in [0,\infty)}$ is a process on a countable space S, the jump times, or arrival times, of X are the random variables $(T_j)_{j \in \mathbb{N}}$ defined by*

$$T_0 = 0, \forall j \geq 1 : \quad T_j := \inf\{t \geq T_{j-1} : X_t \neq X_{T_{j-1}}\}.$$

The waiting times (other names include holding times, sojourn times, and even interarrival times) are defined by $S_j = T_j - T_{j-1}$ if $T_{j-1} < +\infty$, and $S_j = +\infty$ if $T_{j-1} = +\infty$.

The fact that trajectories are right constant has the following implications for the jump times:

Remark 3.4. For every right continuous process $(X_t)_{t \in [0,\infty)}$ on a countable space S, the jump times form a strictly increasing sequence as long as they remain finite:

$$\forall j \geq 1 : \quad P(S_j > 0) = 1.$$

In particular, we have $P(T_1 > 0) = 1$, and as a consequence of Corollary B.3, we have

$$\lim_{h \to 0^+} P(T_1 < h) = 0, \ \lim_{h \to 0^+} P(S_j < h) = 0. \tag{3.1}$$

Since the jump times T_n form an increasing sequence, they will (pointwise, that is, for every ω) have a limit as $n \to \infty$, which may or may not be finite. If this limit is finite, this means that the process undergoes an infinite number of jumps over a finite time interval, a phenomenon called *explosion*:

Definition 3.5. *The explosion time of the process X is the random variable τ_X, defined as*

$$\tau_X := \sup_n T_n = \sum_{n=0}^{\infty} S_n.$$

The process x is said to be nonexplosive if T_n converges almost surely to ∞ as $n \to \infty$, and explosive otherwise.

Strictly speaking, τ_X should be called the *first* explosion time of X; however, we shall never consider a process beyond this point, and we will take care in specifying this by writing $X = (X_t)_{t \in [0, \tau_X)}$.

Many of the continuous-time processes that appear in modeling are integer-valued, and are primarily defined through their increments; here is some relevant vocabulary:

Definition 3.6. *A counting process is a continuous-time process $(X_t)_{t \in [0, \tau_X)}$ taking only nonnegative integer values that is increasing in the sense that for all $s, t \in [0, \tau_X)$ satisfying $s \leq t$ we have $P(X_s \leq X_t) = 1$. A continuous-time process $(X_t)_{t \in [0, \tau_X)}$ is said to have*

1. *stationary increments if for all $s, t \in [0, \tau_X)$ with $t > s$ the distribution of the increment $X_t - X_s$ depends only on the difference $t - s$ (in other words, for all $t, s, h > 0$ with $s < t$ and $t + h \in [0, \tau_X)$, the increments $X_{t+h} - X_{s+h}$ and $X_t - X_s$ have the same distribution: $X_{t+h} - X_{s+h} \stackrel{d}{=} X_t - X_s$);*
2. *independent increments if for every choice of $t_1, \ldots, t_n$ satisfying $0 \leq t_1 < t_2 < \cdots < t_n < \tau_X$ the variables $X_{t_2} - X_{t_1}, \ldots, X_{t_n} - X_{t_{n-1}}$ are independent.*

Note that in some books, stationarity of increments is defined by the following stronger condition involving an arbitrary number of increments:

$$(X_{t_2+h} - X_{t_1+h}, \ldots, X_{t_n+h} - X_{t_{n-1}+h}) \stackrel{d}{=} (X_{t_2} - X_{t_1}, \ldots, X_{t_n} - X_{t_{n-1}}).$$

Let us begin with a result on the first jump time:

Theorem 3.7. *Let $(X_t)_{t>0}$ be a nonexplosive counting process with stationary and independent increments, and assume that $X_0 = 0$. Then its first jump time is exponentially distributed.*

Proof. Since X is increasing for all $t, s > 0$, we have the equality of events

$$\{X_{t+s} = 0\} = \{X_r = 0, \forall r \in [0, t+s]\},$$

and thus

$$P(T_1 > s + t) = P(X_{t+s} = 0) = P(X_{t+s} - X_s = 0 \cap X_s = 0).$$

From the independence and stationarity of the increments we get

$$P(T_1 > s + t) = P(X_{t+s} - X_s = 0)P(X_s = 0)$$
$$= P(X_t = 0)P(X_s = 0) = P(T_1 > t)P(T_1 > s),$$

and the characterization of the exponential distribution provided by Corollary A.5 now tells us that T_1 is exponentially distributed.   □

3.2 The Poisson process

3.2.1 A handmade construction

Perhaps the simplest way to motivate the definition of the Poisson process is to try to generalize the notion of uniform distribution to the case of an unbounded interval, $[0, \infty)$ for instance. For $T > 0$ fixed, let us draw n points uniformly in $[0, T]$. For every $t \in [0, T]$, the number N_t of points falling in the subinterval $[0, t]$ has the binomial distribution $\mathcal{B}(n, \frac{t}{T})$, since each point falls in $[0, t]$ with probability $\frac{t}{T}$. Therefore, the expectation of the number of points in $[0, t]$ is $E(N_t) = n\frac{t}{T}$. Our goal is now to take T and n extremely large while keeping this expectation finite and nonzero; a simple way to do this is to take $\frac{T}{n}$ constant. Therefore, let us define $T = \frac{n}{\lambda}$, with λ a positive constant. Then for all $t > 0$, if n is large enough, the distribution of N_t is $\mathcal{B}(n, \frac{\lambda t}{n})$, and Lemma B.8 tells us that N_t converges in distribution to a Poisson variable X_t of parameter λt. This means that in this asymptotics we have obtained a time-continuous process $(X_t)_{t \geq 0}$, with $X_t \sim \mathcal{P}(\lambda t)$ for each t. How are these variables related? If $t_1 < t_2$, then the increment $X_{t_2} - X_{t_1}$ is obtained as the limit of $N_{t_2} - N_{t_1}$. Therefore, one may expect the limit process X to have stationary and independent increments. These heuristics motivate the formal definition that follows (note that here we implicitly require the process to be nonexplosive):

3.2.2 Formal definition

Definition 3.8 (Poisson process, first definition). *A counting process* $(X_t)_{t>0}$ *is said to be a Poisson process if the following hold:*

1. *it has independent increments;*
2. *it has stationary increments;*
3. $X_0 = 0$, *and the trajectories are right constant;*
4. *there exists a constant* $\lambda > 0$ *such that for all* $t > 0$, *one has* $X_t \sim \mathcal{P}(\lambda t)$.

From stationarity of increments, for $s \leq t$ we obtain $X_t - X_s \sim \mathcal{P}(\lambda(t - s))$, and thus (recall that the expectation of a Poisson variable is its parameter)

$E(X_t - X_s) = \lambda(t - s)$. This means that over a time interval $[s, t]$ the process typically increases by a number that is proportional to the length $t - s$.

The third condition says that as time increases from 0, the process stays at 0 for a certain amount of time until it jumps (meaning jump times do not accumulate near the initial time). The following equivalence tells us how large X_h can be for small h:

$$P(X_h > 1) = 1 - P(X_h = 0) - P(X_h = 1) = 1 - e^{-\lambda h}(1 + \lambda h)$$
$$= 1 - (1 - \lambda h + o(h))(1 + \lambda h) = o(h). \quad (3.2)$$

In other words, the probability that X_h is larger than 1 is vanishingly small as $h \to 0$. Surprisingly, this "little-o condition" suffices to recover the Poisson distribution, i.e., it provides us with an alternative definition of the Poisson process:

Theorem 3.9 (Poisson process, second definition). *A counting process* $(X_t)_{t>0}$ *is a Poisson process if and only if*

1. *it has independent increments;*
2. *it has stationary increments;*
3. $X_0 = 0$, *and the trajectories are right constant;*
4. $P(X_h > 1) = o(h)$ *as* $h \to 0$.

Proof. We have only to show that if X satisfies these four conditions, then there exists a constant $\lambda > 0$ such that $X_t \sim \mathcal{P}(\lambda t)$ for all $t > 0$.

Given that X has stationary and independent increments, our proof lends itself to the use of the Laplace transform. More precisely, dividing the time interval $[0, t]$ into n subintervals of equal length $\frac{t}{n}$, we obtain a representation of X_t as a sum of n independent copies of $X_{\frac{t}{n}}$:

$$X_t = [X_{\frac{t}{n}} - X_0] + \cdots + [X_{\frac{nt}{n}} - X_{\frac{(n-1)t}{n}}].$$

Let $\phi(t)$ denote the Laplace transform of X_t (omitting the z-dependence to have lighter notation):

$$\phi(t) := E(\exp(-zX_t)).$$

Taking the Laplace transform on both sides of the representation of X_t, we obtain

$$\phi(t) = [\phi(\frac{t}{n})]^n.$$

Note that since X is increasing, for all $z > 0$ the function $\phi(\cdot)$ is nonincreasing. We may then apply Lemma A.4 to get

$$\phi(t) = \phi(1)^t.$$

Setting $\mu := -\ln \phi(1)$ (note that $0 \leq \mu < \infty$, since ϕ is positive nondecreasing and thus $0 < \phi(1) \leq 1$), this becomes $\phi(t) = e^{-\mu t}$ for all $t \geq 0$. It follows that ϕ is right differentiable at 0, and $\mu = -\phi'(0)$. Let us now compute this derivative using a difference quotient:

$$\forall h > 0 : \frac{1 - \phi(h)}{h} = \frac{1}{h} E[1 - \exp(-zX_h)] = \frac{1}{h} \sum_{k=0}^{\infty} (1 - e^{-zk}) P(X_h = k)$$

$$= (1 - e^{-z}) \frac{P(X_h = 1)}{h} + \frac{1}{h} \sum_{k=2}^{\infty} (1 - e^{-zk}) P(X_h = k).$$

The last term vanishes in the limit $h \to 0$ thanks to property 4 of X; since $\mu < \infty$, the last right-hand side must have a finite limit as $h \to 0$, so we may conclude that the limit

$$\lambda := \lim_{h \to 0} \frac{P(X_h = 1)}{h}$$

exists and is finite. Thus we have $\mu = (1 - e^{-z})\lambda$, and returning to ϕ, we get the relation

$$E(\exp(-zX_t)) = \exp(-\lambda t(1 - e^{-z})),$$

which exactly characterizes the fact that $X_t \sim \mathcal{P}(\lambda t)$. $\square$

Let us now consider the amplitude of jumps:

Theorem 3.10. *Every Poisson process is simple, which means that all its jumps have magnitude 1:*

$$\forall j \geq 1 : X_{T_j} = j.$$

Proof. We could use either one of our two equivalent definitions to obtain the result; let us use the second one (see Problem 3.1 for a proof based on the first definition). In order to approximate events formulated in terms of the whole process $(X_t)_{t>0}$ by events defined only in terms of finitely many of its values, we discretize the time interval. The idea here is that all jumps will be "seen" on a time grid if the mesh is small enough; to ensure that the amplitudes of the jumps behave in a monotonic fashion in passing to a finer mesh, we use nested grids. More precisely, for fixed $T > 0$ consider the event of a jump of amplitude strictly greater than 1 (call such events "large jumps") before time T:

$$J := \{\exists t \in (0, T] : \quad X_t - X_{t-} > 1\}$$

(here X_{t-} denotes the left-hand limit of X at t). We use the sequence of nested meshes

$$G_n := \{T \frac{k}{2^n}, 0 \leq k \leq 2^n\}$$

and consider the event of a large jump on the grid G_n:

$$J_n := \{\exists k \in \{0, \ldots, 2^n - 1\} \, X_{T\frac{k+1}{2^n}} - X_{T\frac{k}{2^n}} > 1\} = \bigcup_{k=0}^{2^n-1} \{X_{T\frac{k+1}{2^n}} - X_{T\frac{k}{2^n}} > 1\}.$$

The event J is well approximated by J_n in the sense that (J_n) is a decreasing sequence of events such that

$$J = \bigcap_{n=0}^{\infty} J_n;$$

therefore, from Theorem B.1 we have $P(J) = \lim_{n\to\infty} P(J_n)$. The probability of J_n is easy to bound from above using the stationarity of the increments:

$$P(J_n) \le 2^n P(X_{\frac{T}{2^n}} > 1).$$

By condition 4, this is $o(1)$ as $n \to \infty$. Therefore, $P(J) = 0$. $\square$

3.2.3 Jump times

Now that we know that all jumps have magnitude 1, we may guess that knowledge of the distribution of arrival times will suffice for the description of the process, hence our desire to study arrival times. For the particular case of Poisson processes, Theorem 3.7 can be considerably sharpened:

Theorem 3.11. *Let X be a Poisson process with intensity λ, and let T_k denote its arrival times as in Definition 3.3. Then for every n, the density of the vector $(T_1, \ldots, T_n)$ is given by*

$$f_{(T_1,\ldots,T_n)}(t_1,\ldots,t_n) = \lambda^n e^{-\lambda t_n} \mathbb{1}_{t_1 < t_2 < \cdots < t_n}. \tag{3.3}$$

Proof. Let T^n denote the random vector $(T_1, \ldots, T_n)$. Obviously, since the arrival times form a strictly increasing sequence, the probability law of T^n is supported on the set $\{0 < t_1 < t_2 < \cdots < t_n\}$; to find its density we are going to evaluate the probability of the vector T^n falling into a given n-dimensional interval, and then let its size shrink to 0. More precisely, for fixed $0 < t_1 < \cdots < t_n$, choose n numbers $s_1, \ldots, s_n$ such that

$$0 < s_1 < t_1 \le s_2 < t_2 \le \cdots \le s_n < t_n,$$

and let R^n be the n-dimensional interval

$$R^n := (s_1, t_1] \times \cdots \times (s_n, t_n].$$

We compute the probability of the event $\{T^n \in R^n\}$ (if the notational complexity bothers you too much, try to write down this proof for $n = 2$) by reformulating it in terms of the increments of X (making use of Theorem 3.10):

$$\begin{aligned}
P(T^n \in R^n) &= P(s_1 < T_1 \le t_1 \cap s_2 < T_2 \le t_2 \cap \cdots \cap s_n < T_n \le t_n) \\
&= P(X_{s_1} = 0 \cap X_{t_1} = 1 \cap X_{s_2} = 1 \cap X_{t_2} = 2 \cap \cdots \\
&\quad \cap X_{s_{n-1}} = n-2 \cap X_{t_{n-1}} = n-1 \cap X_{s_n} = n-1 \cap X_{t_n} \ge n) \\
&= P(X_{s_1} = 0 \cap X_{t_1} - X_{s_1} = 1 \cap X_{s_2} - X_{t_1} = 0 \cap \cdots \\
&\quad \cap X_{t_{n-1}} - X_{s_{n-1}} = 1 \cap X_{s_n} - X_{t_{n-1}} = 0 \cap X_{t_n} - X_{s_n} \ge 1). \tag{3.4}
\end{aligned}$$

Stationarity and independence of the increments now give us

$$P(T^n \in R^n) = P(X_{t_1} = 0)P(X_{t_1-s_1} = 1)P(X_{s_2-t_1} = 0)\cdots$$
$$P(X_{t_{n-1}-s_{n-1}} = 1)P(X_{s_n-t_{n-1}} = 0)P(X_{t_n-s_n} \geq 1). \quad (3.5)$$

Each probability in the above product is an exponential; after taking into account all exponent cancellations, we get the explicit expression

$$P(T^n \in R^n) = \lambda^{n-1}(t_1 - s_1)\cdots(t_{n-1} - s_{n-1})e^{-\lambda s_n}(1 - e^{-\lambda(t_n-s_n)}).$$

Dividing by the volume $vol(R^n) = (t_1 - s_1)\cdots(t_n - s_n)$, we obtain

$$\frac{P(T^n \in R^n)}{vol(R^n)} = \lambda^{n-1}e^{-\lambda s_n}\frac{1 - e^{-\lambda(t_n-s_n)}}{t_n - s_n}.$$

Letting each s_i converge to t_i, we have obtained the announced density for T^n. $\quad\square$

The above statement was formulated in terms of the arrival times T_k, but it may be reformulated in terms of the interarrival times S_i thanks to the following (you may think of the next result as a general statement about sums of i.i.d. exponential variables, without any reference to a Poisson process):

Theorem 3.12. *Let $(S_j)_{j\geq 1}$ be an infinite collection of continuous positive random variables, and for each $k \geq 1$ set $T_k := S_1 + \cdots + S_k$. Then the density of $(T_1,\ldots,T_n)$ is given by (3.3) for every $n \geq 1$ if and only if the variables $(S_j)_{j\geq 1}$ are i.i.d. exponential variables of parameter λ. In this case, each T_n is a $\Gamma(n,\lambda)$ variable.*

Proof. Necessity first: since $S_k = T_k - T_{k-1}$, the distribution of $S^n := (S_1,\ldots,S_n)$ may be obtained from that of T^n using the method of test functions. To this end, let $\phi : (0,\infty)^n \to \mathbb{R}$ be a smooth function and let us compute the expectation $E(\phi(S^n))$:

$$E(\phi(S^n)) = E(\phi(T_1, T_2 - T_1, \ldots, T_n - T_{n-1}))$$
$$= \int_{t_1=0}^{\infty}\int_{t_2=t_1}^{\infty}\cdots\int_{t_n=t_{n-1}}^{\infty}\phi(t_1, t_2 - t_1, \ldots, t_n - t_{n-1})\lambda^n e^{-\lambda t_n}\,dt_1\cdots dt_n.$$
$$(3.6)$$

Using the change of variable $s_1 = t_1$ and $s_i = t_i - t_{i-1}$ for $i \geq 2$ we obtain the relation

$$E(\phi(S^n)) = \int_{[0,\infty)^n}\phi(s_1, s_2, \ldots, s_n)\lambda^n e^{-\lambda(s_1+\cdots+s_n)}\,ds_1\cdots ds_n.$$

This implies that the density of S^n is given by

$$f_{(S_1,\ldots,S_n)}(s_1, \ldots, s_n) = \lambda^n e^{-\lambda(s_1+\cdots+s_n)},$$

or equivalently, that the variables $S_1, \ldots, S_n$ are i.i.d. exponential variables of parameter λ. The reverse implication is also proved using the method of test functions, by basically reversing the above computation. If ϕ is a test function as before, then

$$E(\phi(T^n)) = E(\phi(S_1, S_1 + S_2, \ldots, S_1 + \cdots + S_n))$$

$$= \int_{s_i > 0} \phi(s_1, s_1 + s_2, \ldots, s_1 + \cdots + s_n) \lambda^n e^{-\lambda(s_1 + \cdots s_n)} \, ds_1 \cdots ds_n, \quad (3.7)$$

and the change of variable $t_i = s_1 + \cdots + s_i$ gives (3.3). Finally, the distribution of T_n is an immediate consequence of this fact (see Appendix B for details if a brushup on the gamma distribution is needed). □

At this point one might think that a simple process whose jump times satisfy the above condition should be a Poisson process: once we know when and by how much the process jumps, what else could there be to specify? In order to design a clean mathematical proof, we need the following technical lemma:

Lemma 3.13. *Let $(S_i)_{i \in \mathbb{N}}$ be a collection of i.i.d. exponential random variables of parameter λ, and for all $k \geq 1$ let $T_k := S_1 + \cdots + S_k$. Define a process $(X_t)_{t \geq 0}$ by $X_0 = 0$ and*

$$\forall t > 0 : \quad X_t := \sum_{k=1}^{\infty} \mathbb{1}_{T_k \leq t}. \quad (3.8)$$

Then conditional on $\{X_t = n\}$, the vector $(T_1, \ldots, T_n)$ has the same distribution as the vector of order statistics of n i.i.d. uniform variables on the interval $[0, t]$:

$$(T_1, \ldots, T_n) | \{X_t = n\} \overset{d}{=} (U_{(1)}, \ldots, U_{(n)}),$$

where the U_i's are i.i.d., and $U_i \sim \mathcal{U}([0, t])$.

Proof. Note that (3.8) says that X is a right constant simple counting process that jumps at precisely the times T_k, $k \geq 1$. This implies the following equalities of events:

$$\{X_t \geq n\} = \{T_n \leq t\}, \quad (3.9)$$

$$\{X_t = n\} = \{T_n \leq t < T_{n+1}\}. \quad (3.10)$$

Let T^n be as in the proof of Theorem 3.11; if I is a given n-dimensional interval, in order to compute the conditional probability $P(T^n \in I | X_t = n)$ we first consider the probability of the intersection, using (3.10) and the distribution of T^{n+1} as given by Theorem 3.11:

$$P(T^n \in I, X_t = n) = P(T^n \in I \cap T_n \leq t < T_{n+1})$$

$$= \int_{I \times \mathbb{R}} \lambda^{n+1} e^{-\lambda t_{n+1}} \mathbb{1}_{0 < t_1 < \cdots < t_n \leq t < t_{n+1}} \, dt_1 \cdots dt_{n+1}. \quad (3.11)$$

Using Fubini's theorem, this integral may be simplified:

$$P(T^n \in I \cap X_t = n) = \int_I (\int_t^\infty \lambda e^{-\lambda t_{n+1}} \, dt_{n+1}) \lambda^n \mathbb{1}_{0<t_1<\cdots<t_n \leq t} \, dt_1 \cdots dt_n$$

$$= \int_I \lambda^n e^{-\lambda t} \mathbb{1}_{0<t_1<\cdots<t_n \leq t} \, dt_1 \cdots dt_n. \quad (3.12)$$

Dividing by $P(X_t = n)$ (recall that $X_t \sim \mathcal{P}(\lambda t)$), we obtain

$$P(T^n \in I | X_t = n) = \int_I \frac{n!}{t^n} \mathbb{1}_{0<t_1<\cdots<t_n \leq t} \, dt_1 \cdots dt_n,$$

which means that conditional on $\{X_t = n\}$, the density of T^n is exactly the integrand in the last integral; in view of Theorem B.9, this is exactly the density of the vector $(U_{(1)}, \ldots, U_{(n)})$. $\square$

We are now in a position to prove the characterization of the Poisson process in terms of its arrival times:

Theorem 3.14 (Poisson process, third definition). *A right constant simple process X satisfying $X_0 = 0$ is a Poisson process of intensity λ if and only if its interarrival times $(S_j)_{j \geq 1}$ form an infinite collection of i.i.d. exponential variables of parameter λ.*

Proof. Necessity has already been proved; let us now show that if the sequence $(S_j)_{j \geq 1}$ is as above, the four conditions of Definition 3.8 are satisfied. We know from Theorem 3.12 that $T_n \sim \Gamma(n, \lambda)$; the distribution of X_t is readily obtained using (3.9):

$$P(X_t \geq n) = P(T_n \leq t) = \int_0^t \lambda^n \frac{s^{n-1}}{(n-1)!} e^{-\lambda s} \, ds.$$

We may use integration by parts (differentiating the term $\lambda^n e^{-\lambda s}$) to get

$$P(X_t \geq n) = [\lambda^n e^{-\lambda s} \frac{s^n}{n!}]_0^t + \int_0^t \frac{s^n}{n!} \lambda^{n+1} e^{-\lambda s} \, ds$$

$$= \frac{(\lambda t)^n}{n!} e^{-\lambda t} + P(X_t \geq n+1). \quad (3.13)$$

Therefore, $X_t \sim \mathcal{P}(\lambda t)$. Now consider the increments of X. Given n points $0 < t_1 \cdots < t_n$, we wish to find the distribution of the vector of increments $(X_{t_1}, X_{t_2} - X_{t_1}, \ldots, X_{t_n} - X_{t_{n-1}})$, i.e., to compute the probability of the event

$$E := (X_{t_1} = k_1 \cap X_{t_2} - X_{t_1} = k_2 \cap \cdots \cap X_{t_n} - X_{t_{n-1}} = k_n)$$

for every choice of n nonnegative integers $k_1, \ldots, k_n$. Let us reformulate E in terms of jump times: setting $r_1 = k_1, r_2 = k_1 + k_2, \ldots, r_n = k_1 + \cdots + k_n$, realization of E means precisely that out of the first r_n jump times, the first k_1

fall in the interval $(0, t_1]$, then the next k_2 jump times fall in the interval $(t_1, t_2]$, and so on until the last k_n jumps fall in the interval $(t_{n-1}, t_n]$. Therefore, the probability of E conditional on $\{X_{t_n} = r_n\}$ is given by

$$P(E|X_{t_n} = r_n) = P(T_1 \in (0, t_1] \cap \cdots \cap T_{k_1} \in (0, t_1] \cap T_{k_1+1} \in (t_1, t_2] \cap \cdots$$
$$\cap T_{r_2} \in (t_1, t_2] \cap \cdots T_{r_n} \in (t_{n-1}, t_n] | X_{t_n} = r_n). \quad (3.14)$$

By Lemma 3.13 this is the probability that if we draw a collection of r_n i.i.d. variables $(U_1, \ldots, U_{r_n})$ all uniform in $[0, t_n]$, for each j between 1 and n, a number k_j of these variables will fall in the interval $(t_{j-1}, t_j]$ (to have homogeneous notation here, we define $t_0 := 0$). Each variable U_l falls into the interval $(t_{j-1}, t_j]$ with probability $\frac{t_j - t_{j-1}}{t_n}$, and thus conditional on $\{X_{t_n} = r_n\}$, the probability of E is given by the multinomial distribution:

$$P(E|X_{t_n} = r_n) = r_n! \prod_{i=1}^n \frac{(t_i - t_{i-1})^{k_i}}{t_n^{k_i}} \frac{1}{k_i!}.$$

Noting the inclusion $E \subset \{X_{t_n} = r_n\}$, we then get

$$P(E) = P(E \cap X_{t_n} = r_n) = P(E|X_{t_n} = r_n)P(X_{t_n} = r_n).$$

Using the fact that $X_{t_n} \sim \mathcal{P}(\lambda t_n)$, this gives an explicit expression for $P(E)$; after some simplification, we get

$$P(E) = \prod_{i=1}^n \frac{[\lambda(t_i - t_{i-1})]^{k_i}}{k_i!} e^{-\lambda(t_i - t_{i-1})}.$$

This implies that the increments are independent and that they are distributed as follows:

$$X_{t_i} - X_{t_{i-1}} \sim \mathcal{P}(\lambda(t_i - t_{i-1})).$$

Finally, this tells us that X has stationary increments: for all $s < t$, the variables $X_t - X_s$ and X_{t-s} have the same distribution $\mathcal{P}(\lambda(t - s))$.   □

Finally, we have a constructive definition of a Poisson process: given an infinite collection $(S_i)_{i \in \mathbb{N}}$ of i.i.d. exponential variables of parameter λ, we set

$$\begin{cases} \text{for} \quad k \geq 1: T_k := S_1 + \cdots + S_k; \\ \text{for} \quad t > 0: X_t := \sum_{k=0}^\infty \mathbb{1}_{T_k \leq t}. \end{cases} \quad (3.15)$$

Theorem 3.14 tells us that a process X is a Poisson process of intensity λ if and only if it is of the form (3.15) (obviously, the process defined by (3.15) is simple and right constant); the T_k's are the jump times of X, and X_t is just the number of jumps in the time interval $[0, t]$; equivalently,

$$X_t = \sup\{k \geq 1 : T_k \leq t\}.$$

3.2.4 The bus paradox

The so-called bus paradox is a subtlety of the Poisson process, or should we say of plainly conditional probability, which basically shows that the time lapse between two consecutive jumps is not as intuitive a notion as one might think. Let us fix $t > 0$, and (assuming that t is not a jump time) consider the two jumps that bracket t. These are respectively T_{X_t} (last jump before t) and T_{X_t+1} (first jump after t). Here are some trivial facts about these two jump times:

Lemma 3.15.

$$X_{T_{X_t}} = X_t;$$
$$\text{for} \quad s \le t : \{T_{X_t} > s\} = \{X_s < X_t\};$$
$$t, s > 0 : \{T_{X_t+1} - t > s\} = \{X_{t+s} < X_t + 1\}.$$

The first equality comes from the fact that the process is right constant; in the second we have two equivalent ways of saying that the last jump before t occurs strictly later than s; finally, in the third we have two equivalent ways of saying that at time $t + s$, jump number $N_t + 1$ has not yet occurred. This has the following unexpected consequence:

$$T_{X_t+1} - t \sim \mathcal{E}(\lambda).$$

Indeed, by stationarity we have

$$P(T_{X_t+1} - t > s) = P(X_{t+s} < X_t + 1) = P(X_s < 1) = e^{-\lambda s}.$$

In order to stick to tradition, let us recast this fact in the language of public transport: if X_t is the number of buses that arrive at a bus stop in the interval $[0, t]$, then $T_{X_t+1} - t$ is the time that remains to wait after time t before a bus arrives. Probabilistic folklore then says that a user who gets to the bus stop at time t will have to wait an average time $E(T_{X_t+1} - t) = \frac{1}{\lambda}$ for the bus, and this is the same as the average time lapse between two arrivals; hence the name "bus paradox." What about the average time elapsed since the last bus arrived? Here is an easy way to compute it based on the second equality in Lemma 3.15:

$$E(T_{X_t}) = \int_0^\infty P(T_{X_t} > t)\, dt = \int_0^t P(T_{X_t} > t)\, dt = \int_0^t P(X_s < X_t)\, dt.$$

Using stationarity and the fact that $X_t - X_s \sim \mathcal{P}(\lambda(t - s))$, you may compute this integral to obtain

$$E(T_{X_t}) = t - \frac{1}{\lambda}(1 - e^{-\lambda t}).$$

For $t > 0$ fixed, let us denote by E_t the elapsed time since the last bus arrived, $E_t := t - T_{X_t}$, and by W_t the waiting time for the next bus, $W_t := T_{X_t+1} - t$. The following picture may help clarify the meaning of these waiting times:

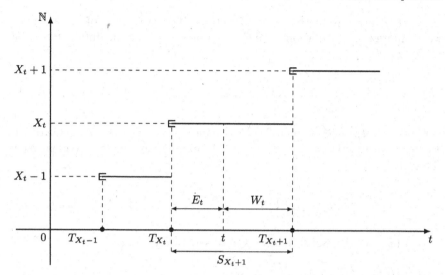

Fig. 3.1. various waiting times involved in the bus paradox

We have obtained above the averages of E_t and W_t:

$$E(E_t) = \frac{1}{\lambda}(1 - e^{-\lambda t}), \quad E(W_t) = \frac{1}{\lambda}.$$

Combining the two, we obtain

$$E(T_{X_t+1} - T_{X_t}) = \frac{1}{\lambda}(2 - e^{-\lambda t}).$$

This is (in some sense) the expected time between two consecutive jumps, although for every k we have $E(T_{k+1} - T_k) = \frac{1}{\lambda}$. Note that we have the strict inequality $E(T_{X_t+1} - T_{X_t}) > \frac{1}{\lambda}$, yet another manifestation of the "bus paradox." Various handwaving explanations may be provided for this strict inequality, but probably it is best to say that in this sort of situation, computations are the best substitute for intuition, which is in fact an excellent a posteriori justification for the development of this whole mathematical machinery. Let us just mention that this is also sometimes called the *inspection paradox*. In this context, inspecting means choosing a time t, and the above inequality is sometimes interpreted as a statement that "inspection increases the gap between jumps."

If we want to (mathematically) investigate the matter further, we may in fact derive the joint distribution of E_t and W_t:

Theorem 3.16. *The variables E_t and W_t are independent, W_t is an exponential random variable of parameter λ, and E_t is an exponential random variable of parameter λ truncated at t.*

Proof. Let us use the method of test functions. If ϕ is a function of two variables, we compute the expectation $E(\phi(E_t, W_t))$ by separating the different possible values for N_t:

$$E(\phi(E_t, W_t)) = \sum_{k=0}^{\infty} E(\phi(E_t, W_t)\mathbb{1}_{N_t=k}) = \sum_{k=0}^{\infty} E(\phi(E_t, W_t)\mathbb{1}_{T_k \leq t < T_k + S_{k+1}}).$$

If $T_k \leq t < T_k + S_{k+1}$, then $T_{N_t} = T_k$ and $T_{N_t+1} = T_k + S_{k+1}$. Therefore, $E_t = t - T_k$ and $W_t = T_k + S_{k+1} - t$. The expectation we wish to compute is therefore

$$E(\phi(E_t, W_t)) = \sum_{k=0}^{\infty} E(\phi(t - T_k, T_k + S_{k+1} - t)\mathbb{1}_{T_k \leq t < T_k + S_{k+1}}).$$

Let us treat the first term (i.e., $k = 0$) separately:

$$E(\phi(t, S_1 - t)\mathbb{1}_{t < S_1})$$
$$= \int_t^{\infty} \phi(t, s - t)\lambda e^{-\lambda s}\, ds = \int_0^{\infty} \phi(t, w)\lambda e^{-\lambda(t+w)}\, dw. \quad (3.16)$$

On the other hand, for $k \geq 1$, using the joint distribution of the pair (T_k, S_{k+1}), we have

$$E(\phi(t - T_k, T_k + S_{k+1} - t)\mathbb{1}_{T_k \leq t < T_k + S_{k+1}})$$
$$= \int_{\tau=0}^{t} \int_{s=t-\tau}^{\infty} \phi(t - \tau, \tau + s - t)\lambda e^{-\lambda \tau}\frac{(\lambda \tau)^{k-1}}{(k-1)!}\lambda e^{-\lambda s}\, ds\, d\tau. \quad (3.17)$$

The change of variable $e = t - \tau, w = \tau + s - t$ gives (after simplification)

$$E(\phi(t - T_k, T_k + S_{k+1} - t)\mathbb{1}_{T_k \leq t < T_k + S_{k+1}})$$
$$= \int_{e=0}^{t} \int_{w=0}^{\infty} \phi(e, w)\lambda^2 e^{-\lambda(t+w)}\frac{(\lambda(t - e))^{k-1}}{(k-1)!}\, dw\, de. \quad (3.18)$$

We may now sum over k:

$$\sum_{k=1}^{\infty} E(\phi(t - T_k, T_k + S_{k+1} - t)\mathbb{1}_{T_k \leq t < T_k + S_{k+1}})$$
$$= \int_{e=0}^{t} \int_{w=0}^{\infty} \phi(e, w)\lambda^2 e^{-\lambda(t+w)} \sum_{k=1}^{\infty} \frac{(\lambda(t - e))^{k-1}}{(k-1)!}\, dw\, de. \quad (3.19)$$

The sum over k is just the exponential $e^{\lambda(t-e)}$; thus grouping the exponents, we obtain

$$\sum_{k=1}^{\infty} E(\phi(t-T_k, T_k+S_{k+1}-t)\mathbb{1}_{T_k \leq t < T_k+S_{k+1}}) = \int_{e=0}^{t} \int_{w=0}^{\infty} \phi(e,w)\lambda^2 e^{-\lambda(e+w)}\, dw\, de.$$

We are almost done! We just need to add the $k=0$ term as we computed it to obtain, finally,

$$E(\phi(E_t, W_t)) = \int_{w=0}^{\infty} \left[e^{-\lambda t}\phi(t,w) + \int_{e=0}^{t} \phi(e,w)\lambda e^{-\lambda e}\, de \right] \lambda e^{-\lambda w}\, dw. \quad (3.20)$$

The cumulative distribution function of E_t may be obtained from (3.20) by taking $\phi(E_t, W_t) := \mathbb{1}_{E_t \leq s}$:

$$P(E_t \leq s) = \int_{w=0}^{\infty} \left[e^{-\lambda t}\mathbb{1}_{t \leq s} + \int_{e=0}^{\min(t,s)} \lambda e^{-\lambda e}\, de \right] \lambda e^{-\lambda w}\, dw.$$

This quantity may be computed exactly, and we obtain

$$P(E_t \leq s) = \begin{cases} 1 & \text{for} \quad s \geq t, \\ 1 - \lambda e^{-\lambda s} & \text{for} \quad s < t, \end{cases}$$

which shows that $E_t \sim T\mathcal{E}(\lambda, t)$ (see Appendix B for details on the truncated exponential distribution). Finally, if we take $\phi(e,w) = \alpha(e)\beta(w)$ in (3.20), we obtain the equality

$$E(\alpha(E_t)\beta(W_t)) = E(\alpha(E_t))E(\beta(W_t))$$

(making use of (B.2)) for every pair of test functions α, β, which proves that that the variables E_t and W_t are independent. □

3.3 Problems

3.1. Using the original definition of the Poisson process, Definition 3.8, adapt the proof of Theorem 3.10 to show that it is a simple process (or see [5] Chapter 8).

3.2. Use (3.3) to derive the density of T_n directly.

3.3. Theorem 3.11 gives the exact density of the vector T^n, and its proof turned out to be relatively technical. If only the distribution of the time T_n is required, a much simpler argument may be used. If a simple process X is such that $X_t \sim P(\lambda t)$ for all $t > 0$, then show by determining the cumulative distribution function of T_n that $T_n \sim \Gamma(n, \lambda)$ (note that such a process is not necessarily a Poisson process; here we make no assumption about the increments).

3.4. In the bus paradox, show directly that $E_t \sim T\mathcal{E}(\lambda; t)$ by determining its cumulative distribution function.

3 The Poisson process

3.5. Let X and Y be two independent Poisson processes, both of intensity λ. Given a Bernoulli variable α independent of both X and Y, set

$$Z := \alpha X + (1 - \alpha)Y.$$

Show that Z is a Poisson process.

3.6. Let X be a Poisson process. Show that for every $n \in \mathbb{N}$ and $0 < s < t$, conditional on $X_t = n$, the variable X_s has the binomial distribution $\mathcal{B}(n, \frac{s}{t})$.

3.7. Let X be a Poisson process of intensity λ, and $(\alpha_n)_{n \in \mathbb{N}}$ an infinite collection of i.i.d. Bernoulli variables of parameter p, all independent of X. Define two processes X^1 and X^2 by

$$\forall t > 0 : \begin{cases} X_t^1 := \sum_{k=1}^{\infty} \mathbb{1}_{T_k \leq t} \mathbb{1}_{\alpha_k = 0}, \\ X_t^2 := \sum_{k=1}^{\infty} \mathbb{1}_{T_k \leq t} \mathbb{1}_{\alpha_k = 1}. \end{cases}$$

This means that every time X jumps, we draw a Bernoulli variable, then increment X^1 by 1 with probability p, and increment X^2 with probability $1 - p$ (you may think of this as arrivals of two different types for the Poisson process X, or arrivals being processed in either one of two different ways). Show that $X_t^1 \sim \mathcal{P}(\lambda p t)$, $X_t^2 \sim \mathcal{P}(\lambda(1 - p)t)$ (in fact, it may be shown that X^1 and X^2 are two independent Poisson processes, called the *split*, or *thinned*, Poisson processes obtained from X). Can you state and prove an analogous result if we want to split a Poisson process into more than two components?

3.8. 1. Compute the characteristic function of a Poisson variable.
   2. Deduce its first two moments and variance.
   3. For this question and the next, let X be a Poisson process of intensity λ. For $s < t$ compute the correlation coefficient of X_t and X_s.
   4. For all $n \in \mathbb{N}$ define Y_n by $Y_n := \frac{X_n}{n}$. Determine the characteristic function Φ_{Y_n} of Y_n and the limit of $\Phi_{Y_n}(z)$ as $n \to \infty$.

3.9. Let $N = (N_t)_{t>0}$ be a Poisson process of intensity λ, and $X_1, \ldots, X_n, \ldots$ an infinite collection of i.i.d. random variables. Define a so-called *compound Poisson process* $(Z_t)_{t>0}$ by $Z_t := X_1 + \cdots + X_{N_t}$. (For instance, N_t is the number of customers coming to a shop in the time interval $[0, t]$, and X_i is the amount spent by the ith customer; the shop manager is interested in the expected value $E(Z_t)$ of sales in the period $[0, t]$.)

   1. Let Φ. denote the characteristic function of a random variable: $\Phi_X(z) := E(\exp(izX))$. Show that

$$\Phi_{Z_t}(x) = e^{-\lambda t} \exp(\lambda t \Phi_{X_1}(z)).$$

   2. Deduce that $E(Z_t) = \lambda t E(X_1)$.

3.10. Here is how to derive a continuous-time version of the Ehrenfest model from the Poisson process. Let us return to the two boxes of Problem 1.11, but this time taking time as a continuous variable. We assume that each ball changes its box after a time that is exponentially distributed, and that all the times of change of box for all balls are i.i.d.; in other words, for each ball the number of times it changes its box is a Poisson process of parameter λ, and all these Poisson processes are independent. Let $N(t)$ denote the number of balls in the left box at time t, and let $p_k(t) := P(N(t) = k)$.

1. Show that for any given ball, the probability to be at time t in the same box as where it started at time 0 is given by

$$p(t) = \frac{1 + e^{-2\lambda t}}{2}.$$

2. Let us now assume that initially all balls are in the left box: $P(N(0) = d) = 1$. Show that in this case, the distribution of $N(t)$ is

$$p_k(t) = \binom{d}{k}\left(\frac{1 + e^{-2\lambda t}}{2}\right)^k \left(\frac{1 - e^{-2\lambda t}}{2}\right)^{d-k}. \tag{3.21}$$

3. Find the limit of $p_k(t)$ as $t \to \infty$; what familiar distribution do you recognize?

4. If we now take the initial condition $N(0) = j$, show that

$$p_k(t) = p(t)^{d-j-k}(1 - p(t))^{j+k} \sum_{l=0}^{k} \binom{j}{l}\binom{d-j}{k-l}\left(\frac{p(t)}{1 - p(t)}\right)^{2l}.$$

5. In this case, what is the limit distribution?

3.11. For $t > 0$ define the sequence of times $(S_k^t)_{k \geq 1}$ by $S_1^t = W_t$ and $S_k^t = S_{X_t+k}$ for $k \geq 1$; in other words, for $k \geq 1$, S_k^t is the time lapse between the $(k-1)$th jump after t and the next one. We are going to show that for every $t > 0$ and $k \geq 1$, the variables $S_1^t, \ldots, S_k^t$ are i.i.d. exponential of parameter λ.

1. Show that for all $s > 0$ and $n \in \mathbb{N}$,

$$P(S_1^t > s | X_t = n) = e^{-\lambda s},$$

and deduce that $S_1^t \sim \mathcal{E}(\lambda)$.

2. Show that for all $s_1 > 0, \ldots, s_k > 0$,

$$P(S_1^t > s_1 \cap \cdots \cap S_k^t > s_k \cap X_t = 0) = e^{-\lambda t} \exp\left(-\lambda(s_1 + \cdots + s_k)\right).$$

3. For all $n \geq 1$ compute the probability

$$P(S_1^t > s_1 \cap \cdots \cap S_k^t > s_k \cap X_t = n);$$

then proceed as in the proof of Theorem 3.16 to deduce that

$$P(S_1^t > s_1 \cap \cdots \cap S_k^t > s_k) = \exp\left(-\lambda(s_1 + \cdots + s_k)\right).$$

4. Conclude.

4

Continuous time, discrete space

Summary. We now move on to general continuous-time Markov processes on countable spaces, i.e., we consider processes $(X_t)_{t>0}$ for which the time variable may assume any value in $[0, \infty)$ (or some bounded interval), and each $X(t)$ takes its values in some countable space $\mathcal{S}$. While the concept of a Markov transition semigroup introduces itself as a generalization of the one-step transition matrix of a discrete-time process, the task of extracting information on the dynamics of the process now becomes significantly harder. The Kolmogorov equations will be derived by essentially analytical methods, which (as before in the case of the Poisson process) rely on dicretizing the process, which means using skeletons. In order to keep the presentation elementary, some of the results we give are not optimal, and whenever necessary, references are given for sharper results.

4.1 Continuous-time Markov chains and their discretizations

4.1.1 The Markov property and the transition function

The definition of a Markov process on $\mathcal{S}$ mimics the discrete-time definition:

Definition 4.1. *Let $\mathcal{S}$ be a countable set. A process $(X_t)_{t>0}$ taking values in $\mathcal{S}$ is said to be a Markov process (or equivalently a Markov chain) on $\mathcal{S}$ if the following conditions are satisfied:*

- *All trajectories $X_t(\cdot)$ are right constant.*
- *For all $s, t > 0$, all integer n, every choice of $n+1$ points in time prior to s, $s_0 < s_1 < \cdots < s_n < s$, and every choice of $n+3$ states $x_0, \ldots, x_n, x, y \in \mathcal{S}$ such that $P(X_s = x \cap X_{s_n} = x_n \cap \cdots \cap X_{s_0} = x_0) \neq 0$, we have*

$$P(X_{t+s} = y | X_s = x \cap X_{s_n} = x_n \cap \cdots \cap X_{s_0} = x_0) = P(X_{t+s} = y | X_s = x).$$

The definition taken here is as close as possible to the one we saw in the discrete-time case; in particular, note that only finitely many times are used in the conditional probability. Since time is now a continuous variable, it is not

© Springer International Publishing AG, part of Springer Nature 2018
J.-F. Collet, *Discrete Stochastic Processes and Applications*,
Universitext, https://doi.org/10.1007/978-3-319-74018-8_4

clear that this will be sufficient to treat events defined in terms of infinitely (possibly uncountably) many times. It will turn out that the first condition is exactly what is needed to ensure that conditioning with only finitely many positions is sufficient to deal with any kind of event formulated in terms of the process X.

As in the discrete-time case, we will consider only processes such that for every pair x, y, the transition probabilities depend only on the time lapse:

Definition 4.2. *A Markov chain $(X_t)_{t>0}$ on a countable space S is said to be homogeneous if for all $x, y \in S$ and all $s, t \geq 0$ we have*

$$P(X_{t+s} = y | X_s = x) = P(X_t = y | X_0 = x).$$

In this case, for all $t \geq 0$, the transition matrix in time t is the Markov matrix p_t defined by

$$p_t(x, y) = P(X_t = y | X_0 = x).$$

Henceforth, we will just say "Markov chain" to mean "homogeneous Markov chain."

Perhaps before venturing into the general theory of continuous-time Markov chains it is worth mentioning that the familiar Poisson process is indeed a Markov process; this turns out to be a consequence of stationarity and independence of its increments:

Theorem 4.3. *Let $X = (X_t)_{t>0}$ be an $\mathbb{N}$-valued process with stationary and independent increments satisfying $X_0 = 0$. Then X is a Markov process on $\mathbb{N}$. Let $(p_t)_{t>0}$ be its transition function; then for all $t > 0$ and $x, y \in \mathbb{N}$:*

$$p_t(x, y) = P(X_t = y - x).$$

Proof. Take $t, s, s_0 < s_1 < \cdots < s_n < s$ as in Definition 4.1. Let us reformulate the conditional probability in Definition 4.1 in terms of the increments of X; in order to do so, first note that the event

$$\{X_s = x \cap X_{s_n} = x_n \cap \cdots \cap X_{s_0} = x_0\}$$

may be rewritten in terms of the increments of the process as the intersection

$$I := \{X_s - X_{s_n} = x - x_n\} \cap \bigcap_{k=1}^{n} \{X_{s_k} - X_{s_{k-1}} = x_k - x_{k-1}\} \cap \{X_{s_0} = x_0\}.$$

The conditional probability in Definition 4.1 now becomes

$$P(X_{t+s} = y | X_s = x \cap X_{s_n} = x_n \cap \cdots \cap X_{s_0} = x_0) = P(X_{t+s} - X_s = y - x | I).$$

Using independence of the increments, this is equal to $P(X_{t+s} - X_s = y - x)$. On the other hand (using independence again), we have

$$P(X_{t+s} = y|X_s = x) = \frac{P(X_{t+s} = y \cap X_s = x)}{P(X_s = x)}$$

$$= \frac{P(X_{t+s} - X_s = y - x \cap X_s = x)}{P(X_s = x)} = P(X_{t+s} - X_s = y - x).$$

This proves that X is a Markov process; then using the stationarity of the increments and the fact that $X_0 \doteq 0$, this last quantity is equal to $P(X_t = y - x)$.
□

As a special case we obtain the transition matrix of a Poisson process of intensity λ (recall that $X_t \sim \mathcal{P}(\lambda t)$):

$$p_t(x, y) = e^{-\lambda t} \frac{(\lambda t)^{y-x}}{(y_x)!}. \tag{4.1}$$

4.1.2 Discretizing a continuous-time Markov chain

In the discrete-time case, knowledge of the transition matrix p_1 suffices for a complete description of the dynamics, since transitions over k units of time are described by the matrix $p_k = p_1^k$. In fact, here we can "sample" the continuous-time process at evenly spaced times to define a discrete-time Markov chain; the following result is immediate:

Theorem 4.4. *Let $X := (X_t)_{t \in [0,\infty)}$ be a continuous-time Markov chain on a countable space $\mathcal{S}$, and let $h > 0$ be some chosen time mesh. Then the discrete-time process $Y := (Y_k)_{k \in \mathbb{N}}$ defined by $Y_k := X_{kh}$ is a discrete-time Markov chain, with one-step transition matrix p_h.*

The process Y will be called the *h-skeleton* of X.

The basic strategy in what follows will be to use a skeleton Y to approximate (in some sense) the process X, which means relating events attached to X (call them "continuous events") to events that are formulated in terms of Y only (call them "discrete events"). More precisely, continuous events need to be approximated from both above and below by discrete events (also note that since the Markov property as we defined it involves only finitely many times, some events involving infinitely many times will require caution).

When applying this strategy to the investigation of the properties of T_1, a key observation is that this approximation may be done if $T_2 - T_1$ is small enough (relative to the mesh h). For instance, for every integer m we have the following inclusion of events:

$$\{X_0 = X_h = \cdots = X_{mh}\} \cap \{T_2 - T_1 > h\} \subset \{X_s = X_0 \ \forall s \in [0, mh]\}.$$

Indeed, if $X_0 = X_h = \cdots = X_{mh} = x$ and X jumps between two grid points kh and $(k+1)h$, then there has to be a second jump before $(k+1)h$ in order to bring us back to x, which means we must have $T_2 - T_1 \leq h$. This yields the following inclusion:

4 Continuous time, discrete space

$$\{X_0 = X_h = \cdots = X_{mh}\} \subset \{X_s = X_0 \quad \forall s \in [0, mh]\} \cup \{T_2 - T_1 < h\}. \quad (4.2)$$

In view of (3.1), this tells us that for small h, the continuous event $\{X_s = X_0 \quad \forall s \in [0, mh]\}$ is well approximated from below by the discrete event $\{X_0 = X_h = \cdots = X_{mh}\}$.

In the investigation of the first jump time T_1 we will make use of the following technical result:

Lemma 4.5. *For every time mesh $h > 0$, every $n \in \mathbb{N}$, and every pair of states $x, y \in S$ with $x \neq y$ we have the following inclusions of events:*

$$\{T_1 > nh\} \subset \{X_0 = X_1 = \cdots = X_{nh}\} \subset \{T_1 > nh\} \cup \{T_2 - T_1 \leq h\}; \quad (4.3)$$

$$\{T_1 \leq nh \cap X_0 = x \cap X_{T_1} = y\}$$
$$\subset \bigcup_{m=1}^{n} \{X_0 = X_h = \cdots = X_{(m-1)h} = x \cap X_{mh} = y\} \cup \{T_2 - T_1 < h\}$$
$$\subset \{T_1 \leq nh \cap X_0 = x \cap X_{T_1} = y\} \cup \{T_2 - T_1 < h\}. \quad (4.4)$$

Proof. The first inclusion in (4.3) is obvious; to prove the second, it suffices to show that

$$\{X_0 = X_1 = \cdots = X_{nh} \cap T_1 \leq nh\} \subset \{T_2 - T_1 \leq h\}.$$

If $X_0 = X_1 = \cdots = X_{nh}$ and $T_1 \leq nh$, then T_1 is not a grid point, meaning that for some $k \leq n - 1$ we have $kh < T_1 < (k+1)h$, and $X_s = X_0$ for all $s \in [0, kh]$. Therefore, we have a (first) jump after kh, followed by a (second) jump some time between T_1 and $(k+1)h$:

$$kh < T_1 < T_2 \leq (k+1)h.$$

This immediately implies $T_2 - T_1 \leq h$, so (4.3) is proved.

·The first inclusion in (4.4) needs to be proved only for the event $\{T_1 \leq nh \cap X_0 = x \cap X_{T_1} = y\}$; in this event, let $m \in \{1, \ldots, n\}$ be such that $(m-1)h \leq T_1 < mh$. Then we have $X_0 = X_h = \cdots = X_{(m-1)h} = x$. If $T_2 - T_1 > h$, then (by the same argument as in the proof of (4.2)) we must have $X_{mh} = y$, which proves the first inclusion in (4.4). To show the second one we have only to show that for every $m \in \{1, \ldots, n\}$ we have

$$\{X_0 = X_h = \cdots = X_{(m-1)h} = x \cap X_{mh} = y \cap T_2 - T_1 > h\}$$
$$\subset \{T_1 \leq nh \cap X_0 = x \cap X_{T_1} = y\}.$$

In the event on the left, since $x \neq y$, there has to be a first jump before mh, so $T_1 \leq mh \leq nh$. Using again the same argument as in the proof of (4.2), we obtain $X_{T_1} = y$. $\square$

4.1.3 The distribution of the first jump time

The Markov property basically says that the system starts afresh after each jump, which leads one to think that the waiting times should also display some form of memory loss. The precise mathematical statement is that the first jump time conditional on some initial state is exponentially distributed:

Theorem 4.6. *If $X = (X_t)_{t>0}$ is a nonexplosive Markov process on a countable space S, then for every initial state $x \in S$ there exists a finite constant $\lambda_x \geq 0$ such that for every time $t > 0$,*

$$P(T_1 > t|X_0 = x) = e^{-\lambda_x t}. \tag{4.5}$$

The state x is absorbing if and only if $\lambda_x = 0$.

Proof. If $t > 0$ is fixed, we may apply (4.3) to $h := \frac{t}{n}$ to obtain

$$P(T_1 > t|X_0 = x) \leq P(X_0 = X_{\frac{t}{n}} = \cdots = X_t|X_0 = x)$$

$$\leq P(T_1 > t|X_0 = x) + P(T_2 - T_1 \leq \frac{t}{n}|X_0 = x).$$

If we now let $n \to \infty$, from (3.1) we obtain the following limit:

$$P(T_1 > t|X_0 = x) = \lim_{n \to \infty} P(X_0 = X_{\frac{t}{n}} = \cdots = X_t|X_0 = x).$$

Finally, using Theorem 4.4, we obtain

$$P(T_1 > t|X_0 = x) = \lim_{n \to \infty} p_{\frac{t}{n}}(x, x)^n. \tag{4.6}$$

Writing $u(s) := p_s(x, x)$ and $\alpha(s) := P(T_1 > s|X_0 = x)$ for all $s > 0$, we have shown that

$$\forall t > 0: \quad \lim_{n \to \infty} u(\frac{t}{n})^n = \alpha(t).$$

Since α is decreasing, we may apply Lemma A.4; thus $\alpha(t) = \alpha(1)^t$ for all $t > 0$, and $\alpha(1)$ must be less than 1. Setting $\lambda_x := -\ln \alpha(1)$, we have $\alpha(t) = e^{-\lambda_x t}$. If $\lambda_x = 0$, then conditional on $X_0 = x$, the jump time T_1 is surely infinite, i.e., x is absorbing; conversely, if x is nonabsorbing, the probability $P(T_1 > t|X_0 = x)$ is strictly less than 1 for t large enough. Therefore, λ_x is positive. Finally, since $P(T_1 > 0|X_0 = x) = 1$, we must have $\alpha(1) \neq 0$, i.e., $\lambda_x < \infty$. $\square$

As an interesting consequence, since an exponential variable may assume any positive value, if the chain is in state x, it can stay there for as long as it pleases. In particular, a transition from x to x is possible over any length of time:

Corollary 4.7. *Under the assumptions of Theorem 4.7, for every x and $t > 0$ we have*

$$p_t(x, x) \geq e^{-\lambda_x t}.$$

Proof. Note that in the case that x is absorbing, both sides of the inequality are 1. From the inclusion of events

$$\{T_1 > t \cap X_0 = x\} \subset \{X_t = x \cap X_0 = x\}$$

we obtain

$$e^{-\lambda_x t} = P(T_1 > t | X_0 = x) \leq P(X_t = x | X_0 = x) = p_t(x, x).$$

$\square$

4.2 The semigroup approach

4.2.1 Markov semigroups

A difficulty appears when one considers an entire continuous process instead of just the sampled chains of the last theorem. We now need to figure out how the various matrices $(p_t)_{t>0}$ are related.

Using (1.1), we have

$$p_{s+t}(x, y) = P(X_{s+t} = y | X_0 = x) = \sum_{z \in S} P(X_{s+t} = y \cap X_s = z | X_0 = x)$$

$$= \sum_{z \in S} P(X_{s+t} = y | X_s = z \cap X_0 = x) P(X_s = z | X_0 = x)$$

$$= \sum_{z \in S} P(X_{s+t} = y | X_s = z) P(X_s = z | X_0 = x)$$

$$= \sum_{z \in S} p_t(z, y) p_s(x, z),$$

which in matrix form is just

$$p_{s+t} = p_s p_t. \tag{4.7}$$

Note that the product $p_s p_t$ is well defined thanks to the fact that both p_s and p_t are Markov matrices. Indeed, when the product is written in components, the partial sums of the involved series are trivially bounded: for every finite subset T of S we have

$$\sum_{z \in T} p_t(x, z) p_s(z, y) \leq \sum_{z \in T} p_t(x, z) \leq 1.$$

When written in components, (4.7) has the following intuitive interpretation: in going from x at time 0 to y at time $t + s$, we may pass by any position z at the intermediate time s, and the corresponding transition probability is obtained by summing over z, each transition being weighted by the corresponding transition probability. Here is the corresponding picture:

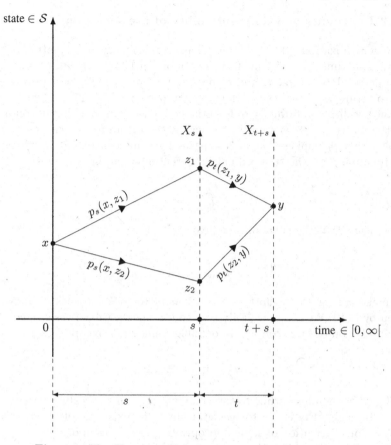

Fig. 4.1. The Chapman–Kolmogorov equation in a picture

We now see that the dynamics of the process are encoded in the map p, which has to satisfy (4.7); hence the following definition:

Definition 4.8. *A transition function (or Markov semigroup) on a countable space S is a map $p : t \mapsto p_t$ defined on $[0, \infty)$ such that*

1. for every $t \geq 0$, p_t is a Markov matrix on S;
2. $p_0 = I$;
3. for every $s, t \geq 0$, the relation (4.7) is satisfied.

As noted above, the first condition implies that the product in (4.7) is well defined; also note that (4.7) implies that the semigroup is commutative: $p_t p_s = p_s p_t$.

4.2.2 Continuity and differentiability of the semigroup

The generalization (4.7) of the Chapman–Kolmogorov equation to the continuous-time case is in fact a doubly infinite system of equations parametrized by s and t, and it does not seem to make it a priori possible to compute p_t for $t > 0$. How can we obtain a Cauchy problem (which means a well-posed ordinary differential equation with some initial condition) for the matrix p_t? If p_t were just a scalar, then from a classical result, (4.7) implies that p_t implies that $p_t = \exp(tq)$ for some constant q, in which case the function p satisfies the ordinary differential equation

$$\frac{dp_t}{dt} = qp_t. \tag{4.8}$$

This means that the constant q is given by

$$q = \frac{dp_t}{dt}_{/t=0^+} = \lim_{h\to 0^+} \frac{p_h - 1}{h}. \tag{4.9}$$

To make a long story short, the solution p_t to (4.7) satisfies (4.8), with q given by (4.9). We can now easily pass to the matrix-valued case: the obvious generalization of (4.9) consists in defining a matrix Q by (provided the limit exists)

$$Q := \lim_{h\to 0^+} \frac{p_h - I}{h}. \tag{4.10}$$

It will take quite a while to turn this into a mathematically rigorous argument; basically, the issues to tackle are first the existence of Q, and second the rigorous derivation of some appropriate generalization of (4.8).

In order to establish the existence of Q, i.e., the differentiability of the semigroup, it turns out that we first need some continuity results. Since we have not been explicit on the notion of convergence in the set of Markov matrices, let us first make precise what we mean by continuity or differentiability of the semigroup:

Definition 4.9. *A Markov semigroup p on a countable set $\mathcal{S}$ is said to be right continuous at 0 if for every pair $x, y \in \mathcal{S}$ the map $p.(x,y)$ is right continuous at 0; similarly, p is said to be continuous on $[0,\infty)$ if it is right continuous at 0 and for every pair $x, y \in \mathcal{S}$ the map $p.(x,y)$ is continuous on $(0,\infty)$. Right differentiability at 0 and differentiability on $[0,\infty)$ are defined analogously.*

Our first regularity result shows that right continuity at 0 is all we need:

Theorem 4.10. *Let $\mathcal{S}$ be a countable space. If a Markov semigroup p on $\mathcal{S}$ is right continuous at 0, then it is continuous on $[0,\infty)$.*

Proof. Let us first treat right continuity at some fixed $t > 0$. For every $h > 0$ and $x, y \in \mathcal{S}$ we have

$$p_{t+h}(x,y) - p_t(x,y) = ((p_h - I)p_t)(x,y) = \sum_{z \in S}(p_h(x,z) - \delta_{x,z})p_t(z,y).$$

Taking absolute values and leaving aside the x term in the sum, we obtain

$$|p_{t+h}(x,y) - p_t(x,y)| \leq \sum_{z \neq x} p_h(x,z)p_t(z,y) + (1 - p_h(x,x))p_t(x,y)$$

$$\leq \sum_{z \neq x} p_h(x,z) + (1 - p_h(x,x))p_t(x,y)$$

$$= (1 - p_h(x,x))(1 + p_t(x,y)).$$

This converges to 0 as $h \to 0$, which shows that $p_\cdot(x,y)$ is right continuous at t. To show left continuity at $t > 0$, choose $0 \leq h \leq t$. Proceeding in a similar fashion, we may write

$$|p_{t-h}(x,y) - p_t(x,y)| = |((I - p_h)p_{t-h})(x,y)|$$

$$\leq \sum_{z \in S}(p_h(x,z) - \delta_{x,z})p_{t-h}(z,y)$$

$$= \sum_{z \neq x} p_h(x,z)p_{t-h}(z,y) + (1 - p_h(x,x))p_{t-h}(x,y)$$

$$\leq 2(1 - p_h(x,x)).$$

This gives left continuity, so finally, $p_\cdot(x,y)$ is continuous at t. $\square$

Note that our proof shows in fact a little more than the original statement: if some x is such that $p_h(x,x) \to 1$ as $h \to 0$, then $p_\cdot(x,y)$ is continuous on $(0,\infty)$ for every y. In other words, right continuity of the diagonal elements at time 0 implies continuity of the whole semigroup at every positive time.

The next issue to consider is differentiability, and a key result is that for a Markov semigroup, continuity ensures differentiability at 0 in the following sense:

Theorem 4.11. *Let $(p_t)_{t>0}$ be a continuous Markov semigroup on a countable space S. Then the matrix $Q := \frac{dp_t}{dt}/_{t=0^+}$ is well defined; more precisely,*

1. For every $x \in S$, the limit

$$Q(x,x) := \lim_{h \to 0^+} \frac{p_h(x,x) - 1}{h}$$

exists and is equal to $-\lambda_x$, where λ_x is the constant of Theorem 4.6.
2. For every $x,y \in S$ with $x \neq y$, the limit

$$Q(x,y) := \lim_{h \to 0^+} \frac{p_h(x,y)}{h}$$

exists.

Proof. Let us begin with the differentiability of $p.(x, x)$. Clearly, if x is absorbing, then $p_t(x, x) = 1$ for all t, and the result is true since $\lambda_x = 0$. If x is not absorbing, in the course of the proof of Theorem 4.6 (see equation (4.6)) we established the following limit:

$$\lim_{n \to \infty} p_{\frac{t}{n}}(x, x)^n = e^{-\lambda_x t}. \qquad (4.11)$$

Since $p.(x, x)$ is right continuous at 0, we have $\lim_{n \to \infty} p_{\frac{t}{n}}(x, x) = 1$, whence the equivalent

$$\ln[p_{\frac{t}{n}}(x, x)^n] = n \ln p_{\frac{t}{n}}(x, x) \sim n(p_{\frac{t}{n}}(x, x) - 1).$$

As $n \to \infty$ this gives the convergence

$$n(p_{\frac{t}{n}}(x, x) - 1) \to -\lambda_x t,$$

and dividing by t yields

$$\forall t > 0 : \frac{p_{\frac{t}{n}}(x, x) - 1}{\frac{t}{n}} \to -\lambda_x. \qquad (4.12)$$

The conclusion now follows immediately from Lemma A.2. To treat the case of $p_t(x, y)$ with $x \neq y$, we may assume that x is not absorbing (the limit is trivially 0 if x is absorbing); let us proceed as in the proof of Theorem 4.6 and apply (4.4). Noting that

$$P(\bigcup_{m=1}^{n} \{X_0 = X_h = \cdots = X_{(m-1)h} = x \cap X_{mh} = y\}|X_0 = x)$$

$$= \sum_{m=1}^{n} p_h(x, x)^{m-1} p_h(x, y) = p_h(x, y) \frac{1 - p_h(x, x)^n}{1 - p_h(x, x)},$$

we obtain

$$P(T_1 \leq nh \cap X_{T_1} = y|X_0 = x)$$

$$\leq p_h(x, y) \frac{1 - p_h(x, x)^n}{1 - p_h(x, x)} + P(S_2 < h|X_0 = x)$$

$$\leq P(T_1 \leq nh \cap X_{T_1} = y|X_0 = x) + P(S_2 < h|X_0 = x).$$

For fixed $t > 0$ we may apply this inequality to $h := \frac{t}{n}$:

$$P(T_1 \leq t \cap X_{T_1} = y|X_0 = x)$$

$$\leq p_{\frac{t}{n}}(x, y) \frac{1 - p_{\frac{t}{n}}(x, x)^n}{1 - p_{\frac{t}{n}}(x, x)} + P(S_2 < \frac{t}{n}|X_0 = x)$$

$$\leq P(T_1 \leq t \cap X_{T_1} = y|X_0 = x) + P(S_2 < \frac{t}{n}|X_0 = x).$$

We may now let $n \to \infty$; in these asymptotics we know from (3.1) that the term involving S_2 vanishes. Therefore,

$$\lim_{n \to \infty} [p_{\frac{t}{n}}(x,y) \frac{1 - p_{\frac{t}{n}}(x,x)^n}{1 - p_{\frac{t}{n}}(x,x)}] = P(T_1 \leq t \cap X_{T_1} = y | X_0 = x).$$

On the other hand, from (4.11) and (4.12) as $n \to \infty$ we have

$$\frac{1 - p_{\frac{t}{n}}(x,x)^n}{1 - p_{\frac{t}{n}}(x,x)} \sim \frac{1 - e^{-\lambda_x t}}{\lambda_x \frac{t}{n}},$$

whence the limit

$$\lim_{n \to \infty} [\frac{p_{\frac{t}{n}}(x,y)}{\frac{t}{n}}] = \frac{\lambda_x}{1 - e^{-\lambda_x t}} P(T_1 \leq t \cap X_{T_1} = y | X_0 = x).$$

The right-hand side is a right continuous function of t (see Corollary B.2), and thus we may apply Lemma A.3. This tells us that the right-hand side does not depend on t (call it $Q(x,y)$) and that it is indeed the right derivative of $p_{.}(x,y)$ at 0:

$$P(T_1 \leq t \cap X_{T_1} = y | X_0 = x) = \frac{1 - e^{-\lambda_x t}}{\lambda_x} Q(x,y), \qquad (4.13)$$

$$\lim_{h \to 0^+} \frac{p_h(x,y)}{h} = Q(x,y). \qquad (4.14)$$

□

An immediate consequence of Theorem 4.11 is the following inequality, which we will use later in the proof of the Kolmogorov forward equation:

Corollary 4.12. *For every $x, y \in S$ with $y \neq x$ and $t > 0$,*

$$\frac{p_t(x,y)}{t} \leq \frac{1 - p_t(x,x)}{t} \leq -Q(x,x). \qquad (4.15)$$

Proof. For every $s, t \geq 0$, the Chapman-Kolmogorov equation gives the inequality

$$p_{t+s}(x,x) \geq p_t(x,x) p_s(x,x).$$

We may therefore use Lemma A.6 to obtain

$$p_t(x,x) \geq 1 + t Q(x,x),$$

whence the second inequality. The first one follows from the fact that since p_t is a Markov matrix, we have $p_t(x,y) \leq 1 - p_t(x,x)$. □

Besides proving differentiability of the semigroup at 0, equation (4.13) immediately yields the distribution of the process immediately after the first jump:

Corollary 4.13. *Let x be any nonabsorbing state of a nonexplosive Markov process X; then*

$$P(T_1 \leq t \cap X_{T_1} = y | X_0 = x)$$
$$= P(X_{T_1} = y | X_0 = x) P(T_1 \leq t | X_0 = x); \quad (4.16)$$

$$\forall y \neq x : \ P(X_{T_1} = y | X_0 = x) = \frac{Q(x, y)}{\lambda_x}. \quad (4.17)$$

To put the statement into words: conditional on $X_0 = x$, the variables X_{T_1} and T_1 are independent, and X_{T_1} is distributed on $S \setminus \{x\}$ according to the probability vector $\frac{Q(x, \cdot)}{\lambda_x}$ (also recall that from (4.5), conditional on $X_0 = x$ the variable T_1 is exponentially distributed with parameter λ_x).

Proof. Combining (4.13) and (4.5), we obtain

$$P(T_1 \leq t \cap X_{T_1} = y | X_0 = x) = \frac{Q(x, \cdot)}{\lambda_x} P(T_1 \leq t | X_0 = x).$$

We may now let $t \to \infty$; since s is nonabsorbing, we may apply Corollary B.4 to get (4.17) and then (4.16). $\square$

4.2.3 The rate matrix and Kolmogorov's equation

For all nonabsorbing states x, relation (4.17) gives

$$\forall y \neq x : \ P(X_{T_1} = y | X_0 = x) = -\frac{Q(x, y)}{Q(x, x)}.$$

Summing over $y \neq x$ and getting rid of the denominator, we obtain the relation

$$Q(x, x) + \sum_{y \neq x} Q(x, y) = 0.$$

Of course, this relation could be formally obtained by differentiating the relation

$$\sum_{y \in S} p_t(x, y) = 0,$$

but if S is not finite, the interchange of the summation and differentiation operations might be tricky to justify. Matrices satisfying such a property deserve to be given a name:

Definition 4.14. *A rate matrix (other names include intensity matrix, Q-matrix, and generator matrix) on a countable space S is a map $Q : S \times S \to \mathbb{R}$ such that $Q(x, y) \geq 0$ for all x, y with $x \neq y$, and*

$$\forall x : \quad \sum_{y \in S} Q(x, y) = 0.$$

In other words, all off-diagonal elements are nonnegative, and all row sums are 0. Note that these two conditions imply that all diagonal elements are nonpositive:

$$\forall x : Q(x, x) \leq 0.$$

We can now restate Theorem 4.11 and some of its consequences as the following:

Theorem 4.15. *For a continuous Markov semigroup* $(p_t)_{t>0}$ *on a countable space* S, *the matrix*

$$Q := (\frac{dp_t}{dt})_{/t=0^+}$$

is well defined, and it is a rate matrix. A state x is absorbing if and only if $Q(x, x) = 0$, or equivalently if and only if $Q(x, y) = 0$ for all y.

The matrix Q will be called the rate matrix, or the infinitesimal generator of the process. Let us look at the particular case of a Poisson process of intensity λ. We know that the transition function is given by (4.1), so the semigroup is obviously continuous; taking the derivative with respect to t and evaluating at $t = 0$, we see that the only terms present in row x of the generator Q are the following:

$$Q(x, x) = -\lambda, \ Q(x, x + 1) = \lambda. \tag{4.18}$$

For readers who like pictures better than formulas,

$$Q = \begin{pmatrix} -\lambda & \lambda & & (0) \\ & -\lambda & \lambda & \\ & & \ddots & \ddots \\ (0) & & & \ddots & \ddots \end{pmatrix}.$$

For general Markov processes, Theorem 4.15 associates to every Markov transition function a rate matrix Q; conversely, if we give ourselves a rate matrix Q, does this define a Markov semigroup? The answer is yes on a finite space:

Theorem 4.16. *An $n \times n$ matrix Q is a rate matrix if and only if for every $t > 0$, the matrix e^{tA} is Markov.*

This result basically says that the dynamics of the process will now be encoded in the matrix A, a much simpler object than the transition function p.

Proof. We make use of the limit

$$e^{tQ} = \lim_{n \to \infty} (I + \frac{tQ}{n})^n.$$

For $n \in \mathbb{N}$ large enough, the matrix $I + \frac{tQ}{n}$ is positive and all its row sums are 1; these properties are preserved in the limit $n \to \infty$, and therefore e^{tQ} is Markov for all $t > 0$. Conversely, if e^{tQ} is Markov for all $t > 0$, we have

$$\forall x : \sum_y (e^{tQ})(x, y) = 1.$$

Taking the derivative with respect to t, we obtain

$$\forall x : \sum_y (Qe^{tQ})(x, y) = 0,$$

and taking $t = 0$, we see that all row sums of Q vanish. We still need to show that all off-diagonal elements of q are nonnegative. Using $e^{tQ} \sim I + tQ$ as $t \to 0^+$, we obtain for $x \neq y$ (you should think for a minute here why it would be incorrect to write $e^{tQ}(x, y) \sim tQ(x, y)$) the limit $\frac{1}{t} e^{tQ}(x, y) \to Q(x, y)$, which gives $Q(x, y) \geq 0$. $\square$

Just as Theorem 4.10 showed that the semigroup property (4.7) propagates continuity, we are now in a position to show that it propagates differentiability. The resulting differential equations are the so-called *Kolmogorov equations*; let us begin with the easy case of a finite space:

Theorem 4.17. *Let $(p_t)_{t>0}$ be a continuous Markov semigroup on a finite space S with rate matrix Q. Then for all $t > 0$, the time derivative of p_t at t exists and satisfies*

$$\frac{dp_t}{dt} = Qp_t, \tag{4.19}$$

$$\frac{dp_t}{dt} = p_t Q. \tag{4.20}$$

Proof. The classical theory of ordinary differential equations (see, for instance, [14]) tells us that when supplemented with the initial condition $p_0 = I$, equations (4.19) and (4.20) give two well-posed Cauchy problems. These two Cauchy problems are equivalent, since their unique solution is given by $p_t = e^{tQ}$ (see the section on matrix exponentials in Appendix C for details), so we just need to establish (4.19).

For every pair of states x, y we may use (4.7) to express $p_{t+h}(x, y)$ and compute a difference quotient:

$$\frac{1}{h}[p_{t+h}(x, y) - p_t(x, y)] = \frac{1}{h} \sum_{z \in S} p_h(x, z)[p_t(z, y) - p_t(x, y)]$$

$$= \frac{1}{h} \sum_{z \in S} p_t(x, z)[p_h(z, y) - p_t(x, y)] \tag{4.21}$$

(we have used the fact that the row sums are 1 for p_h and p_t). Letting h go to 0^+ immediately yields both differential equations. $\square$

If the matrix p_t is to be determined from Q, the computation of the matrix exponential e^{tQ} is in most cases intractable, so the most common method will be to solve either one of the two differential equations; it will turn out that sometimes one of the two may be easier to solve.

What happens if the space $\mathcal{S}$ is infinite? Obviously, the interchange of the summation and the limit $h \to 0^+$ in (4.21) will need to be justified; more precisely, some kind of uniform control of the partial sums has to be derived. Let us begin with a generalization of (4.19):

Theorem 4.18. *Let $(p_t)_{t>0}$ be a continuous Markov semigroup on a countable space $\mathcal{S}$ with rate matrix Q. Then p_t is differentiable on $[0, \infty)$, and for all $t > 0$, we have*

$$\frac{dp_t}{dt} = Qp_t. \tag{4.22}$$

Proof. The idea is to somehow "propagate" right differentiability in time using the semigroup and controlling the infinite sum by a finite sum for which the interchange of limit and summation is allowed. For $x, y \in \mathcal{S}$ fixed, we are going to show that $p_t(x, y)$ is a differentiable function of t on $[0, \infty)$ by showing that the right-hand derivative exists everywhere on $[0, \infty)$ and is continuous (an elementary result in calculus asserts that if a map has a continuous right-hand derivative everywhere on some open interval I, then it is of class $\mathcal{C}^1$ on I). To evaluate the right-hand derivative of $p.(x, y)$ at t, we compute a difference quotient using the relation $p_{t+h} = p_h p_t$ and singling out the diagonal term $p_h(x, x)$:

$$\frac{p_{t+h}(x, y) - p_t(x, y)}{h} = \frac{p_h(x, x) - 1}{h} p_t(x, y) + \sum_{z \neq x} \frac{p_h(x, z)}{h} p_t(z, y).$$

In the right-hand side above, as $h \to 0^+$ the first term converges to $Q(x, x)p_t(x, y)$, which depends continuously on t. Therefore, we need to treat only the sum. More precisely, we are going to show that

$$\lim_{h \to 0^+} \sum_{z \neq x} \frac{p_h(x, z)}{h} p_t(z, y) = \sum_{z \neq x} Q(x, z)p_t(z, y). \tag{4.23}$$

We need to check that the right-hand side of (4.23) depends continuously on t. If $\{z_1, z_2, \dots\}$ is an enumeration of $\mathcal{S}$, then its partial sums are bounded in absolute value as follows:

$$\sum_{i=1, x \neq z_i}^{n} |Q(x, z_i)p_t(z_i, y)| \leq \sum_{i=1, x \neq z_i}^{\infty} |Q(x, z_i)| = |Q(x, x)|$$

(here we used the fact that Q is a rate matrix). By dominated convergence, it follows that the right-hand side of (4.23) is indeed a continuous function of t, so we are left with establishing (4.23).

In order to approximate the infinite sum by a finite one, choose N large enough to ensure that $x \in \{z_1, \dots, z_N\}$. Then we have

$$\sum_{i=1,x\neq z_i}^{N} \frac{p_h(x,z_i)}{h} p_t(z_i,y) \leq \sum_{z\neq x} \frac{p_h(x,z)}{h} p_t(z,y)$$

$$\leq \sum_{i=1,x\neq z_i}^{N} \frac{p_h(x,z_i)}{h} p_t(z_i,y) + \sum_{k>N} \frac{p_h(x,z_k)}{h}. \quad (4.24)$$

Using the fact that p_t is a Markov matrix, the infinite sum above may be reformulated as a finite one:

$$\sum_{k>N} p_h(x,z_k) = 1 - \sum_{j=1}^{N} p_h(x,z_j) = 1 - p_h(x,x) - \sum_{j=1,x\neq z_j}^{N} p_h(x,z_j).$$

In order to simplify the notation, let us define S_h and S_h^N by

$$S_h := \sum_{z\neq x} \frac{p_h(x,z)}{h} p_t(z,y), \quad S_h^N := \sum_{i=1,x\neq z_i}^{N} \frac{p_h(x,z_i)}{h} p_t(z_i,y).$$

Returning to (4.24), we obtain

$$S_h^N \leq S_h \leq S_h^N + \frac{1-p_h(x,x)}{h} - \sum_{j=1,z_j\neq x}^{N} \frac{p_h(x,z_j)}{h}.$$

Letting $h \to 0^+$, we obtain for all N large enough that

$$\sum_{i=1,x\neq z_i}^{N} Q(x,z_i)p_t(z_i,y) \leq \liminf_{h\to 0^+} S_h$$

$$\leq \limsup_{h\to 0^+} S_h \leq \sum_{i=1,x\neq z_i}^{N} Q(x,z_i)p_t(z_i,y) - \sum_{i=1}^{N} Q(x,z_i). \quad (4.25)$$

We may now let $N \to \infty$; in this limit, the last sum involving Q vanishes (since Q is a rate matrix), and we obtain (4.23). $\square$

If we put together the statements of Theorems 4.10, 4.15, and 4.18, we see under no other assumption than right continuity of the semigroup at 0 that the semigroup is differentiable, its right derivative Q at 0 is a rate matrix, and equation (4.22) is satisfied for all $t > 0$. For the generalization of (4.20) it turns out that an extra additional assumption is needed to control the infinite sums:

Theorem 4.19. *Let $(p_t)_{t>0}$ be a continuous Markov semigroup on a countable space S with rate matrix Q. Assume that*

$$K := \sup_{z \in S} |Q(z, z)| < \infty. \tag{4.26}$$

Then p_t is differentiable on $[0, \infty)$, and for all $t \geq 0$, we have

$$\frac{dp_t}{dt} = p_t Q. \tag{4.27}$$

Proof. We begin as in the previous proof, this time using $p_{t+h} = p_t p_h$ and singling out the y term in the sum:

$$\frac{p_{t+h}(x, y) - p_t(x, y)}{h} = \sum_{z \neq y} p_t(x, z) \frac{p_h(z, y)}{h} + \left(\frac{p_h(y, y) - 1}{h} \right) p_t(x, y).$$

As before, we need to show only that this right-hand side has a limit as $h \to 0^+$, and that the limit depends continuously on t. This is obviously true of the final term on the right-hand side (which converges to $p_t(x, y)Q(y, y)$), so we need to treat only the sum. As $h \to 0^+$ the general term converges to $p_t(x, z)Q(z, y)$, and (from (4.15)) it is bounded by $Kp_t(x, z)$. Thus we may apply the dominated convergence theorem for series (Theorem A.8) to conclude that

$$\lim_{h \to 0^+} \frac{p_{t+h}(x, y) - p_t(x, y)}{h} = \sum_{z in S} p_t(x, z)Q(z, y).$$

This quantity depends continuously on t (again by dominated convergence), which completes the proof. $\square$

Theorem 4.20. *Let $(p_t)_{t>0}$ be a continuous Markov semigroup on a countable space S with rate matrix Q; let u_0 and v_0 respectively be a row vector and a column vector with bounded components:*

$$\sup_{x \in S} |u(x)| < \infty, \ \sup_{x \in S} |v(x)| < \infty.$$

For $t \geq 0$ set

$$u_t := u_0 p_t, \ v_t := p_t v_0.$$

Then the following hold:

1. v_t is differentiable on $[0, \infty)$, and for all $t \geq 0$ we have

$$\frac{dv_t}{dt} = Q v_t. \tag{4.28}$$

2. If, moreover, condition (4.26) is satisfied, then u_t is differentiable on $[0, \infty)$, and for all $t \geq 0$ we have

$$\frac{du_t}{dt} = u_t Q. \tag{4.29}$$

4 Continuous time, discrete space

Remark: the boundedness condition on u_0 and v_0 is there only to ensure that the products $u p_t$ and $p_t v$ are well defined; in fact, we shall use (4.28) and (4.29) only for probability vectors.

Proof. If S is finite, then (4.28) and (4.29) are immediate consequences of (4.22) and (4.27) (just multiply by u_0 or v_0); in the general case, (4.28) is proved by repeating the proof of (4.22), and (4.29) by repeating that of (4.27). □

A remark on terminology: equations (4.22) and (4.27) are respectively called the *backward* and *forward Kolmogorov equations*. Here is why: in components, the definition of v_t and equation (4.28) read

$$v_t(x) = \sum_{y \in S} p_t(x, y) v_0(y),$$

$$\frac{d}{dt} v_t(x) = \sum_{y \in S} Q_t(x, y) v_0(y).$$

Recall that $p_t(x, y) = P(X_t = y | X_0 = x)$; in words, this means that in the term $p_t(x, y)$, the point x is where we come from, whence the name *backward*. In a symmetric fashion, the definition of u_t reads

$$u_t(x) = \sum_{y \in S} u_0(y) p_t(y, x).$$

In the term $p_t(y, x)$, the point x is now a destination, whence the name *forward*. By analogy, equations (4.22) and (4.27) are respectively called the backward and forward Kolmogorov equations (another trick for remembering this is to look at the position of the generator Q in the right-hand side of the equation: in (4.22) it is in the back, whereas in (4.27) it is in the front—provided we agree that back means left and front means right). Here is how to give probabilistic substance to Theorem 4.20:

Proposition 4.21. *Let $(p_t)_{t>0}$ be a continuous Markov semigroup on a countable space S with rate matrix Q. Let $g : S \to \mathbb{R}$ be a bounded function, and for $t \geq 0, x \in S$ set*

$$\mu(t, x) := P(X_t = x), \quad u(t, x) := E(g(X_t) | X_0 = x).$$

Then the following hold:

1. For all $t \geq 0$,

$$\frac{\partial u}{\partial t}(t, x) = (Qu)(t, x). \tag{4.30}$$

2. If (4.26) is satisfied, then for all $t \geq 0$,

$$\frac{\partial \mu}{\partial t}(t, x) = (\mu Q)(t, x). \tag{4.31}$$

92

Again, these two equations are called (respectively) backward and forward equations; the forward equation (4.31) is also sometimes called the *Fokker–Planck equation* (the greek letter (in contrast to v above) μ is chosen here in anticipation of the case of continuous spaces, where the corresponding object will be a measure, while (4.30) will become a partial differential equation). In modeling situations it is quite usual to define a process by specifying a rate matrix and using either Kolmogorov (backward or forward) equation to deduce information about p_t.

4.3 Communication classes

4.3.1 The embedded chain

The sequence $Y := (Y_n)_{n\in\mathbb{N}}$ of the values assumed by X_t is given by $Y_n = X_{T_n}$. Although it seems rather intuitive that this should be a discrete-time Markov process, the rigorous proof of this fact is quite involved and requires the use of measure theory, a tool far beyond the scope of this text (see, for instance, [5], Theorem 4.2, p. 348, or [26], Theorem 2.8.2, p. 94). The process Y is called the *embedded chain*, or *jump chain* of X (not to be confused with a skeleton of X). If we admit that Y is Markovian, the results of the previous section provide its transition matrix:

Theorem 4.22. *Assume that the nonexplosive Markov process X has a continuous semigroup, hence a generator Q; then the transition matrix A of its embedded chain is given by*

$$\begin{cases} \text{if } \quad Q(x,x) = 0 : \begin{cases} A(x,x) = 1, \\ A(x,y) = 0 \quad \text{for} \quad y \neq x; \end{cases} \\ \text{if } \quad Q(x,x) \neq 0 : \begin{cases} A(x,x) = 0, \\ A(x,y) = -\frac{Q(x,y)}{Q(x,x)} \quad \text{for} \quad y \neq x. \end{cases} \end{cases} \tag{4.32}$$

Proof. The absorbing case $Q(x,x) = 0$ is obvious; if x is not absorbing, i.e., if the constant $\lambda_x = -Q(x,x)$ in Theorem 4.6 is nonzero, then for all $y \neq x$, equation (4.17) yields (since Y is Markovian)

$$E(x,y) = P(X_{T_1} = y|X_0 = x) = -\frac{Q(x,y)}{Q(x,x)}.$$

Since Q is a Markov matrix and Q is a rate matrix, the diagonal element of A is now given by

$$A(x,x) = 1 - \sum_{y\neq x} A(x,y) = 1 + \frac{\sum_{y\neq x} Q(x,y)}{Q(x,x)} = 0.$$

$\square$

Henceforth the matrix A will be called Q's *jump matrix*:

Definition 4.23. *The jump matrix of a rate matrix Q is the matrix A defined by (4.32).*

4.3.2 Communication and recurrence

Accessibility and communication are defined exactly as in the discrete-time case:

Definition 4.24. *State y is said to be accessible from state x if $p_t(x, y) > 0$ for some $t > 0$; also, we shall say that x and y communicate if each is accessible from the other.*

If state y is accessible from state x for a chain X, we write $x \xrightarrow{X} y$, or simply $x \to y$ if the process X is implicit. As in the discrete-time case, communication is an equivalence, and the equivalence classes are called the *communication classes* of the process. If we denote by $(Y_n)_{n \in \mathbb{N}}$ the embedded chain, then the notion of accessibility is the same for X and for Y,

$$x \xrightarrow{X} y \iff x \xrightarrow{Y} y,$$

in which case we shall simply write $x \to y$. In particular, the communication classes are the same for X and Y, and the partition of the state space into disjoint communication classes provided by Theorem 1.26 is the same for both X and Y. As before, the chain X is said to be irreducible if it has only one communication class, which is equivalent to irreducibility of the embedded chain Y. The definitions of recurrence and positive recurrence are again the same as in the discrete-time case:

Definition 4.25. *Assume X is a nonexplosive Markov process; for each state x define the variable T_x by*

$$T_x := \inf\{t > 0 : \quad X_t = x\}.$$

Then x is said to be recurrent if $P(T_x < \infty | X_0 = x) = 1$, and transient otherwise. A recurrent state x is said to be positive recurrent if $E(T_x | X_0 = x) < \infty$, and null recurrent otherwise.

It is easy to see that a state is recurrent for a chain X if and only if it is recurrent for the embedded chain Y. Therefore, as in the discrete case, within a communicating class all points are of the same nature, recurrent or transient. However, positive recurrence for X is not equivalent to that for Y. In view of the convergence results of Chapter 1, we might want to look for some generalization of the notion of aperiodicity; however, the next result shows that there is no analogue in the case of continuous time:

Theorem 4.26. *Let $(p_t)_{t>0}$ be a continuous Markov semigroup, $X = (X_t)_{t>0}$ an associated chain, and A its jump matrix. Then for every nonabsorbing state x, the following hold:*

1. *If $A(x, y) > 0$, then $p_t(x, y) > 0$ for all $t > 0$;*
2. *if $x \to y$, then $p_t(x, y) > 0$ for all $t > 0$;*

in particular, if the chain is irreducible, then $p_t(x, y) > 0$ for all $x, y \in \mathcal{S}, t > 0$.

In other words, y is accessible from x in some time t if and only if it is accessible in any length of time, which is determined by the corresponding element in the jump matrix. This should be rather intuitive, since by Theorem 4.6, a first jump from x to y happens after an exponential time, which can assume any positive value as long as x is not absorbing.

Proof. Fix a nonabsorbing point x, and $y \neq x$ such that $A(x, y) > 0$. From Theorems 4.11 and 4.22, as $h \to 0$ we have

$$p_h(x, y) \sim hQ(x, y) = -hQ(x, x)A(x, y) > 0.$$

Therefore, for some $h_0 > 0$ we have

$$\forall h \in (0, h_0): \quad p_h(x, y) > 0.$$

This says that it is possible to go from x to y in time x; on the other hand, once in y, we may stay there for as long as we please, since Corollary 4.7 tells us that

$$\forall s > 0: \quad p_s(y, y) > 0.$$

Taking $h < h_0$ and $t > h$, we obtain

$$p_t(x, y) \geq p_h(x, y)p_{t-h}(y, y) > 0,$$

which completes the proof of point 1. Now if $x \to y$ (recall that the accessible points from x are the same for both chains X and Y), for some $k \in \mathbb{N}$ we have $A^k(x, y) > 0$. If $k = 1$, we are done (by the first point); if $k \geq 2$, there is a sequence of points $x_1, \ldots, x_{k-1}$ such that

$$A(x, x_1)A(x_1, x_2) \cdots A(x_{k-1}, y) > 0.$$

Using the first part of our conclusion, for every $s > 0$ we have

$$p_s(x, x_1)p_s(x_1, x_2) \cdots p_s(x_{k-1}, y) > 0,$$

and composing paths, we obtain $p_t(x, y) > 0$ for all $t > 0$ (take, for instance, $s = \frac{t}{k}$ to split the trajectory from x to y in t time units into k equal steps). $\square$

4.4 Stationary distributions and convergence to equilibrium

By analogy with the discrete-time case, stationary distributions are defined as probability distributions that are invariant under the evolution on every time interval:

Definition 4.27. *A row vector u is said to be invariant for a Markov semi-group $(p_t)_{t \geq 0}$ if for every $t > 0$ we have $u = up_t$; a stationary distribution is a probability vector π that is invariant.*

In the finite-space case, the notion of invariant vector is not really relevant, because every nonzero invariant vector may be normalized to produce a stationary distribution, so existence of a nonzero invariant vector is equivalent to existence of a stationary distribution. When the state space is infinite, it is quite possible to have an invariant vector u that is not normalizable, i.e., such that $\sum_x u(x)$ is infinite; this is, in particular, the case for some random walks; see the next chapter. Note that if it exists, a stationary distribution π will also be a stationary distribution for every skeleton $(X_{nh})_{n \in \mathbb{N}}$. In contrast to the discrete-time case, the definition involves infinitely many conditions, which say that π is invariant for the evolution on *every* time interval; a more tractable criterion is formulated in terms of the rate matrix:

Proposition 4.28. *Assume that the Markov semigroup $(p_t)_{t \geq 0}$ is continuous, so that it has a generator Q. Assume, moreover, that condition (4.26) is satisfied. Then a probability vector π is a stationary distribution if and only if it satisfies $\pi Q = 0$.*

Proof. If $\pi Q = 0$, we may use (4.29) to obtain

$$\frac{d}{dt}(\pi p_t) = \pi p_t Q = \pi Q p_t = 0.$$

Therefore, the vector πp_t is constant, which (evaluating it at time $t = 0$) gives $\pi p_t = \pi$. Conversely, if πp_t is constant, we obtain $\pi Q p_t = 0$ for all t; thus (taking $t = 0$) we obtain $\pi Q = 0$. $\square$

Remark: if we do not assume the boundedness condition (4.26), the result remains true, provided the rate matrix is irreducible and all states are recurrent; see, for instance, [26], Theorem 3.5.5, p. 120 for a proof.

As in the discrete-time case, it may be the case that transitions from one state to another are balanced pairwise and not just globally; more precisely, detailed balance equilibria are defined as follows:

Definition 4.29. *A probability vector π is a detailed balance equilibrium if for every pair x, y we have $\pi(y)Q(y, x) = \pi(x)Q(x, y)$.*

4.4   Stationary distributions and convergence to equilibrium

Obviously, every detailed balance equilibrium is a stationary distribution (a direct consequence of the fact that Q is a rate matrix); a chain is said to be *reversible* if it has a detailed balance equilibrium. We shall refer to the relations defining a detailed balance equilibrium as the *local* or *detailed balance relations*, as opposed to the *global balance relations*

$$\sum_x \pi(x)Q(x,y) = 0.$$

We noted that a stationary distribution π, if present, also is stationary for every skeleton, and one could expect some connection with the jump chain. The precise result is the following:

Proposition 4.30. *Let A be the jump matrix of a rate matrix Q; a vector π satisfies $\pi A = 0$ if and only if the vector μ defined by $\mu(x) = -Q(x,x)\pi(x)$ satisfies $\mu A = \mu$.*

Proof. If μ is defined in terms of π, then we may compute the components of $\mu(A-I)$ using (4.32):

$$(\mu(A-I))(y) = \sum_{x \in \mathcal{S}} \mu(x)A(x,y) - \mu(y)$$

$$= \sum_{x \neq y, Q(x,x) \neq 0} \pi(x)Q(x,y) + Q(y,y)\pi(y)$$

$$= \sum_{x \in \mathcal{S}} \pi(x)Q(x,y) = (\pi Q)(y)$$

(here we used the fact that if $Q(x,x) = 0$, then $Q(x,y) = 0$ for all y). This means that $\mu(A-I) = \pi Q$, which proves the result. $\square$

Note that this result is not exactly a statement about stationary distributions: here we do not assume that π is a probability vector. If we normalize π so as to make it a probability vector, then the corresponding vector μ will generally not be a probability vector. Here is another way to say this: the definition of μ in terms of π does not map probability vectors to probability vectors. The multiplication by $Q(x,x)$ may, however, be understood intuitively as follows: since we know that the first jump time T_x from a position x has exponential distribution of parameter $-Q(x,x)$, its expectation is $\frac{1}{-Q(x,x)}$. On the other hand, for the jump chain, this time increment is 1; this means that the time scale for the jump chain is locally (when starting from x) distorted by a factor $Q(x,x)$. Note that this handwaving argument could have been used earlier to give a (probably equally unconvincing) heuristic justification for relation (4.32): for x nonabsorbing, everywhere not on the diagonal divide the x row of Q by $-Q(x,x)$ to obtain the x row of A.

If a stationary distribution exists and the chain is irreducible, the convergence result is very similar to (and in fact is deduced from) the discrete-time case:

Theorem 4.31. *Assume that the semigroup $(p_t)_{t>0}$ is continuous, is irreducible, and has a stationary distribution π. Then there is no other stationary distribution, and for all $x, y \in S$ we have*

$$\lim_{t\to\infty} p_t(x, y) = \pi(y).$$

Proof. By Theorem 4.26, for all $t > 0$ all entries of the matrix p_t are positive; in particular, p_1 is a positive matrix. This means that the skeleton $(X_n)_{n\in\mathbb{N}}$ is an irreducible aperiodic chain, and since it has π as a stationary distribution, we may apply Corollary 1.43. This shows that there may be no stationary distribution for X other than π, since any such distribution would also be a stationary distribution for the skeleton; also, we obtain the following convergence:

$$\forall x \in S: \quad \lim_{n\to\infty} \sum_{y\in S} |p_n(x, y) - \pi(y)| = 0.$$

We want to show that this convergence stills holds if n is replaced by $t > 0$. For $t > 0$, define its fractional part $\{t\}$ by $\{t\} := t - [t]$, where $[t]$ is the integer part of t. Then $p_t = p_{[t]}p_{\{t\}}$, whence (using $\pi p_{\{t\}} = \pi$)

$$p_t(x, y) - \pi(y) = \sum_{z\in S} [p_{[t]}(x, z) - \pi(z)]p_{\{t\}}(z, y).$$

The upper bound

$$|p_t(x, y) - \pi(y)| \le \sum_{z\in S} |p_{[t]}(x, z) - \pi(z)|$$

immediately yields the conclusion. $\square$

4.5 Defining a continuous-time Markov chain by its generator

In applications it is quite common to define a process by specifying its rate matrix Q. The issue is then to determine whether such a process exists and what its explosion time would be. In order to define a process with rate matrix Q we need to determine its embedded chain, i.e. (in probability), the sequence of its values at jump times, and its *clock*, i.e., the sequence of its interarrival times $(S_n)_{n\in\mathbb{N}}$. For Q a given rate matrix on S, let us indicate how this construction can be made; for $x \in S$ set $\lambda(x) := -Q(x, x)$. The jump matrix A is necessarily given in terms of Q by (4.32); let us denote by $Z = (Z_n)_{n\in\mathbb{N}}$ the associated discrete-time Markov chain. The clock will now be chosen so as to ensure that conditional on $\{Z_{n-1} = z\}$, the waiting time S_n is $\mathcal{E}(\lambda(z))$. To this end, let us take an infinite collection $(E_i)_{i\in\mathbb{N}}$ of i.i.d. exponential variables, all $\mathcal{E}(1)$ and independent of the process Z. The waiting times S_n and jump times $(T_n)_{n\in\mathbb{N}}$ are now defined as

4.5 Defining a continuous-time Markov chain by its generator

$$S_n := \frac{E_n}{\lambda(Z_{n-1})}, \quad T_n := S_0 + \cdots + S_{n-1}.$$

Note that from Proposition B.5 we have

$$S_n|\{Z_{n-1} = z\} \sim \mathcal{E}(\lambda(z)). \tag{4.33}$$

Finally, X_t is defined from the conditions that its embedded chain is Z, and its interarrival times are the S_i's:

$$X_t := \sum_{n=0}^{\infty} Z_n \mathbb{1}_{[T_n, T_{n+1}[}(t).$$

Let us recap this construction into a neat algorithm:

1. Define A by (4.32).
2. Use A to simulate a discrete-time Markov chain Z.
3. Take the variables $(E_i)_{i \in \mathbb{N}}$ as above, and set

$$\begin{cases} T_n := \frac{E_0}{\lambda(Z_0)} + \cdots + \frac{E_{n-1}}{\lambda(Z_{n-1})}; \\ \text{For} \quad t \in [T_n, T_{n+1}) : \quad X_t = Z_n. \end{cases} \tag{4.34}$$

At this point it is important to realize that our construction gives a single process X with rate matrix Q, but there might be others. In fact, it is possible to show that every process having Q as a rate matrix is of this form (the missing link here is that we have not shown that if a process has Q as a rate matrix, then its interarrival times satisfy (4.33). For a proof, see, for instance, [26], Theorem 2.8.4, p. 97).

This construction makes it easy to address explosion:

Theorem 4.32. *If (4.26) is satisfied, then the process defined by (4.34) is nonexplosive.*

Proof. If (4.26) is satisfied, then

$$T_n \geq \frac{1}{K}(E_1 + \cdots + E_{n-1}),$$

and from Lemma B.6 we have

$$T_n \overset{a.s.}{\to} \infty \quad \text{as} \quad n \to \infty.$$

$\square$

If mathematical rigor is no concern, then one might just go ahead with the determination of a stationary distribution (solving $\pi Q = 0$) and the study of Kolmogorov's equations; in easy cases either one of these will be exactly solvable, yielding an exact expression for p_t.

4.6 Problems

4.1. Let S_r be the class of square matrices whose row sums are all equal to r. Show that $A \in S_r$ and $B \in S_s$ implies $AB \in S_{rs}$, that S_0 is closed under addition, and finally that $B \in S_0$ implies $AB \in S_0$ for every square matrix A (or in algebraic parlance, that S_0 is a *left ideal* in the ring of square matrices). In particular, this shows that both S_0 and S_1 are closed under matrix multiplication; however, the product of two rate matrices is not necessarily a rate matrix, and in fact, even the square of a rate matrix is not necessarily a rate matrix. Can you check this on a simple 2×2 example?

4.2. Show that for every transition function p and $h > 0$, the quantity $\sum_j |p_{ij}(t+h) - p_{ij}(t)|$ is a nonincreasing function of $t \geq 0$.

4.3. If X is a Poisson process of intensity, then for all $t \geq 0$ set $Y_t = (-1)^{X_t}$; show that Y is a Markov process and determine its transition function. This is called the *telegraph* or *flip-flop* process.

4.4. Consider a general rate matrix A and transition matrix $p - t$ written as

$$A := \begin{pmatrix} -\lambda & \lambda \\ \mu & -\mu \end{pmatrix}, p_t := \begin{pmatrix} u(t) & 1 - u(t) \\ 1 - v(t) & v(t) \end{pmatrix}.$$

1. Show that Kolmogorov's forward and backward equations respectively take the form

$$\begin{cases} u'(t) = -\lambda(u+v) + \lambda, \\ v'(t) = -\mu(u+v) + \mu, \end{cases} \quad \begin{cases} u'(t) = -(\lambda+\mu)u + \mu, \\ v'(t) = -(\lambda+\mu)v + \lambda. \end{cases}$$

2. Show that if one takes the same initial condition $p(0)$, then both problems are equivalent if and only if the following condition is satisfied:

$$\lambda(1 - v(0)) = \mu(1 - u(0)).$$

Hint: this is a direct consequence of a result in the appendix on calculus, or it may be shown by transforming the first system.

4.5. Let A be the 2×2 jump matrix given by $A(1,2) = \lambda$, $A(2,1) = \mu$, and let p_t denote the associated transition matrix.

1. What is the associated equilibrium distribution?
2. Is it a detailed balance equilibrium?
3. Show by solving the Kolmogorov equation $p'_t = p_t A$ that

$$p_t(1,1) = \frac{\mu}{\lambda+\mu} + \frac{\lambda}{\lambda+\mu} \exp(-(\lambda+\mu)t),$$

and compute the other entries of p_t.

4. Show that
$$\lim_{n \to \infty} ((p_{\frac{t}{n}})(1,1))^n = \exp(-\lambda t),$$

and deduce the distribution of the waiting time in state 1.

4.6. In this problem we return to the proof of Theorem 4.18 and derive a continuity result for some functions defined by series. Let $(u_n)_{n \in \mathbb{N}}$ be a collection of maps defined on $(0,1]$ satisfying the following conditions:

$$\forall n \in \mathbb{N}, t \in (0,1] : \quad u_n(t) \geq 0,$$
$$\forall n \in \mathbb{N} : \lim_{t \to 0^+} u_n(t) = w_n,$$
$$\lim_{t \to 0^+} \sum_{n=0}^{\infty} u_n(t) = \sum_{n=0}^{\infty} w_n < \infty.$$

Show that for every bounded nonnegative sequence $(v_n)_{n \in \mathbb{N}}$ we have

$$\lim_{t \to 0^+} \sum_{n=0}^{\infty} u_n(t) v_n = \sum_{n=0}^{\infty} v_n w_n.$$

In other words, if each u_n has a limit at 0 and the series $\sum u_n$ also has such a limit, then for every v as above, the series $\sum v_n u_n$ may also be extended by continuity at 0.

5

Examples

Summary. We shall now take a much more applied point of view, and present some illustrative examples for the general theory of the previous chapters. The emphasis will be on computations rather than mathematical issues such as summability and differentiability. As is often the case in modeling, discrete-time processes will be defined by specifying a one-step transition matrix, and continuous-time process by specifying a generator; we shall implement this approach for respectively random walks and birth–death models. For more examples of processes in scientific modeling we recommend the monograph [11], and for the particular case of random walks the interested reader should consult [19].

5.1 Random walks

The general picture of a discrete-time random walk is quite an intuitive one: in some kind of discrete space an object may move from one point to a restricted number of other points, which will be called its *neighbors*. One of the simplest spaces in which there is such a notion of neighbor is the set of integers, in which the neighbors of k are $k-1$ and $k+1$ (incidentally, $\mathbb{N}$ is not as simple, because the origin is somewhat different from the rest of the points); another case occurs when the space is a directed graph and the neighbors of x are the points y such that (x, y) is an edge. This is slightly reminiscent of what we saw about random surfers in Chapter 2, but in this modest book we shall not consider random walks on graphs. For more reading on random walks, an excellent presentation may be found in [19].

5.1.1 Random walks on $\mathbb{Z}$

Let us begin with the easy case of a random walk on $\mathbb{Z}$. Let $(Z_n)_{n \in \mathbb{N}}$ be an infinite collection of i.i.d. variables taking values in $\{-1, 1\}$ with $P(Z_n = 1) = p$, and X_0 a random variable taking values in $\mathbb{Z}$, independent of the Z_n's. For $n \geq 1$ put $X_n = X_{n-1} + Z_n$, or more explicitly,

© Springer International Publishing AG, part of Springer Nature 2018
J.-F. Collet, *Discrete Stochastic Processes and Applications*,
Universitext, https://doi.org/10.1007/978-3-319-74018-8_5

5 Examples

$$X_n := X_0 + Z_1 + \cdots + Z_n.$$

The process $(X_n)_{n \in \mathbb{N}}$ is called a random walk of parameter p on $\mathbb{Z}$. By linearity, the expected value of X_n is

$$E(X_n) = E(X_0) + n(2p - 1).$$

This means that as n increases we have $E(X_n) \to \infty$ if $p > \frac{1}{2}$, and $E(X_n) \to -\infty$ if $p < \frac{1}{2}$. This is very intuitive: if $p > \frac{1}{2}$, the point X_n has a tendency to move to the right more often than to the left, hence a general drift toward the right. If we take $X_0 = 0$, this leads us to conjecture that 0 will not be revisited much unless $p = \frac{1}{2}$ (in which case $E(X_n) = 0$; the random walk is then said to be *symmetric*). More precisely, this is the only case in which the process is recurrent:

Theorem 5.1. *For all p, a random walk of parameter p on $\mathbb{Z}$ is irreducible and has no stationary distribution. It is transient if $p \neq \frac{1}{2}$, and null-recurrent if $p = \frac{1}{2}$.*

Proof. In each row the transition matrix has 0 on the diagonal, p to the right, and $1 - p$ to the left. Starting from a point x, the process may reach $x - 1$ or $x + 1$ after 1 step, so every point y may be reached in $|y - x|$ steps. This means that the chain is irreducible (however, it is trivially not ergodic). An invariant vector π satisfies

$$\pi_k = (1 - p)\pi_{k+1} + p\pi_{k-1},$$

and therefore

$$\pi_{k+1} - \pi_k = \frac{p}{1 - p}(\pi_k - \pi_{k-1}), \quad k \in \mathbb{Z}.$$

This gives us a two-parameter family of invariant vectors. Indeed, solving this recurrence relation, we obtain the expression

$$\pi_k = \pi_1 + \left[\frac{p}{1 - p} + \cdots + \left(\frac{p}{1 - p}\right)^{k-1}\right](\pi_1 - \pi_0), \quad |k| > 1, \qquad (5.1)$$

in which π_0, π_1 can be any two constants. If $\pi_1 \neq \pi_0$, then the components π_k are unbounded (recall that here $k \in \mathbb{Z}$), and if $\pi_1 = \pi_0$, they are all equal; in either case (leaving aside the trivial case $\pi = 0$) their sum is infinite, which proves that there is no stationary distribution. Theorem 1.34 tells us that there can be no positive recurrent state, so the chain is either null recurrent or transient. To settle this issue we may use Proposition 1.20 and compute $p^n(0,0)$ to check the nature of the point 0 (now that we know that all points are of the same nature). If n is odd, then obviously the chain cannot go from 0 to 0 in n steps (note that X_n changes parity each time), and thus $p^n(0,0) = 0$. If n is even, say $n = 2k$, then going from 0 to 0 in $2k$ steps means taking k steps to the right and k steps to the left exactly. Such a path will henceforth be called a *2k-loop at* 0. For a $2k$-loop, the number of steps to the right (or in more mathematical parlance the number of indices l such that $Z_l = +1$) follows the binomial distribution $\mathcal{B}(2k, p)$; thus

$$p^{2k}(0,0) = \binom{2k}{k} p^k (1-p)^{n-k}.$$

In order to find an equivalent as $k \to \infty$, we may use Stirling's formula (A.7); after some elementary manipulation, we obtain

$$p^{2k}(0,0) \sim [4p(1-p)]^n \frac{1}{\sqrt{\pi n}} \quad as \quad k \to \infty.$$

Proposition 1.20 tells us that the point 0 is transient if and only if the series $\sum_k p^{2k}(0,0)$ is finite; for $p \neq \frac{1}{2}$ we have $4p(1-p) < 1$, and thus by the ratio test this series converges, and all points are transient. Finally, if $p = \frac{1}{2}$, then $p^{2k}(0,0) \sim \frac{1}{\sqrt{\pi k}}$. Thus the series diverges, and all points are recurrent. $\quad\square$

5.1.2 Symmetric random walks on $\mathbb{Z}^d$

The generalization to $\mathbb{Z}$ of the symmetric one-dimensional random walk is very easy to define: the immediate neighbors of a point $x \in \mathbb{Z}^d$ are the points obtained by incrementing one coordinate of x by ± 1. Note that there are $2d$ such neighbors; in other words, the random walk will jump only parallel to the coordinate axes, and the magnitude of jumps is always 1. If we call $e_1, \ldots, e_d$ the standard basis vectors of $\mathbb{Z}^d$, mathematically this means that we are giving ourselves an infinite collection of i.i.d. random vectors $(Z_n)_{n \in \mathbb{N}}, Z_n \in \mathbb{Z}^d$ with $P(Z_n = e_i) = P(Z_n = -e_i) = \frac{1}{2d}$, and as before, we define X_n by $X_n - X_{n-1} = Z_n$. The fundamental recurrence result proved by Pólya in 1922 is the following:

Theorem 5.2 (Pólya's theorem). *The symmetric random walk on $\mathbb{Z}^d$ is recurrent if and only if $d \leq 2$.*

The main tool for the proof is the recurrence criterion of Proposition 1.20. Since the chain is clearly irreducible, we need to check recurrence (or lack thereof) only for the point 0, which means that the issue is the nature of the series $\sum p^n(0,0)$. As in the one-dimensional case, return to the origin is possible only after an even number of steps, so what we need to investigate is the nature of the series $\sum p^{2n}(0,0)$. We are going to examine two different methods for settling this issue, one combinatorial and one based on Fourier analysis. Combinatorics first:

Proof. The case $d = 1$ has been treated before; let us examine the case $d = 2$ as a warmup. We are interested in $2n$-loops at 0. The total number of paths of length $2n$ is 4^{2n}, and for such a path to be a loop it has to take a certain number of steps n_1 to the right (in mathematical terms, the number of increments assuming the value $+e_1$ is n_1) and an equal number of steps to the left.

105

5 Examples

Similarly, it will go up n_2 times and down n_2 times, with the constraint on the total number of steps $2n_1 + 2n_2 = 2n$, i.e., $n_1 + n_2 = n$. Choosing where these four values of increments $\pm e_1, \pm e_2$ are going to be means distributing $2n$ objects over 4 boxes (with the constraint that boxes 1 and 3 contain the same number of objects, boxes 2 and 4 contain the same number of objects); in other words, we are looking at a uniform multinomial distribution. Therefore, $p^{2n}(0,0)$ is given by

$$p^{2n}(0,0) = (\frac{1}{4})^{2n} \sum_{n_1+n_2=n} \frac{(2n)!}{(n_1!n_2!)^2}.$$

This double sum can be easily transformed into a simple sum to determine an equivalent:

$$p^{2n}(0,0) = (\frac{1}{4})^{2n} \sum_{k=0}^{n} \frac{(2n)!}{(k!(n-k)!)^2} = (\frac{1}{4})^{2n} \binom{2n}{n} \sum_{k=0}^{n} \binom{n}{k}^2.$$

Using Vandermonde's identity (B.3), we obtain

$$p^{2n}(0,0) = (\frac{1}{4})^{2n} \binom{2n}{n}^2.$$

Now Stirling's formula (A.7) finishes the job for us: after some elementary algebra you will find that as $n \to \infty$,

$$p^{2n}(0,0) \sim \frac{1}{\pi n}.$$

This means that the series $\sum p^{2n}(0,0)$ diverges, and thus the chain is recurrent. In higher dimension $d > 2$, the structure is pretty much the same, and by the same argument (you should by now have a good feeling for how a $2n$-loop is built) we obtain

$$p^{2n}(0,0) = (\frac{1}{2d})^{2n} \sum_{n_1+\cdots+n_d=n} \frac{(2n)!}{n_1!\cdots n_d!} = (\frac{1}{2d})^{2n} \binom{2n}{n} \sum_{n_1+\cdots+n_d=n} (\frac{n!}{n_1!\cdots n_d!})^2.$$

From the multinomial formula we have

$$\sum_{n_1+\cdots+n_d=n} \frac{n!}{n_1!\cdots n_d!} = (1+\cdots+1)^n = d^n; \tag{5.2}$$

in order to find an upper bound for the general term of this sum we use inequality (A.8):

$$\min_{n_1+\cdots+n_d=n} n_1!\cdots n_d! \geq [\Gamma(1+\frac{n}{d})]^d.$$

Inserting in the expression for $p^{2n}(0,0)$, we obtain the upper bound

106

$$p^{2n}(0,0) \le (\frac{1}{2d})^{2n} \binom{2n}{n} \frac{n!}{[\Gamma(1+\frac{n}{d})]^d}$$

(in which we used (5.2)). Using Stirling's formula (A.7), you will find this upper bound to be equivalent to $\sqrt{2}(\frac{d}{2\pi})^{\frac{d}{2}} \frac{1}{n^{\frac{d}{2}}}$; this shows for $d > 2$ that the series $\sum_n p^{2n}(0,0)$ is finite; hence the chain is nonrecurrent. $\square$

Let us now look at the Fourier-analytic approach to proving Theorem 5.2.

Proof. Since the position at time n is the sum of n i.i.d. increments, the probability density of X_n may be computed by convolution. Increments being stationary for all $x, y \in \mathbb{Z}$, we have $p^n(x,y) = p^n(0, y-x)$. Calling this quantity $h_n(y-x)$, from independence of the increments, we obtain

$$p^n(x,y) = \sum_{x_1,\dots,x_{n-1} \in \mathbb{Z}^d} p(x,x_1) \cdots p(x_{n-1}, y)$$

$$= \sum_{x_1,\dots,x_{n-1} \in \mathbb{Z}^d} h_1(x_1 - x) \cdots h_1(y - x_{n-1}), \quad (5.3)$$

i.e., $h_n = h_1 * \cdots * h_1$. The Fourier transform $\hat{h}_1^n$ of the map $h_n : \mathbb{Z}^d \to \mathbb{R}$ is therefore given by $\hat{h}_n = \hat{h}_1^n$, whence the need to compute $\hat{h}_1$:

$$z \in [-\pi, \pi]^d : \quad \hat{h}_1(z) = \sum_{k \in \mathbb{Z}^d} e^{-ikz} h_1(k).$$

The one-step transition probability $h_1(k)$ is equal to $\frac{1}{2d}$ if k is an immediate neighbor of 0 and vanishes elsewhere; therefore, we immediately obtain

$$\hat{h}_1(z) = \frac{1}{d} \sum_{j=1}^d \cos z_j.$$

We may now recover h_n using the inversion formula:

$$x \in \mathbb{Z}^d : \quad h_n(x) = (\frac{1}{2\pi})^d \int_{[-\pi,\pi]^d} (\hat{h}_1(z))^n e^{izx} \, dz.$$

Taking $x = 0$ and using the formula for the sum of a geometric series, we obtain

$$\sum_{n=0}^{\infty} p^n(0,0) = (\frac{1}{2\pi})^d \int_{[-\pi,\pi]^d} \frac{1}{1 - \hat{h}_1(z)} \, dz.$$

The issue is now to determine whether this integral is finite or infinite; the only singularity of the integrand is at 0, where (use $\cos h \sim 1 - \frac{h^2}{2}$ near 0) we have the equivalent $\hat{h}_1(z) \sim 1 - |z|^2$, in which $|\cdot|$ is the Euclidean norm. This means that near $z = 0$ the integrand is equivalent to $\frac{1}{|z|^2}$. To make a long story short, our series has a finite sum if and only if $\frac{1}{|z|^2}$ is integrable near 0, which (use spherical coordinates) is true if and only if $d > 2$. $\square$

5 Examples

5.1.3 The Reflecting walk on $\mathbb{N}$

The reflecting walk on $\mathbb{N}$ is the one-sided version of the previous random one-dimensional walk, in which we impose a reflection condition at 0:

$$p(0,0) = 1 - p, \quad p(0,1) = p.$$

Clearly, the chain is irreducible, and since $p(0,0) = 1 - p$ is positive, it is aperiodic. Since the chain is irreducible, we have only to determine the nature of the point 0. This will be a consequence of the exact expression of the hitting probabilities for the point 0:

Theorem 5.3. *For the reflecting walk on $\mathbb{N}$, for all $k \geq 1$,*

$$P(T_0 < \infty | X_0 = k) = \begin{cases} 1 & \text{if} \quad p \leq \frac{1}{2}, \\ \left(\frac{1-p}{p}\right)^k & \text{if} \quad p \geq \frac{1}{2}. \end{cases} \tag{5.4}$$

Proof. Let us begin with the case $k = 1$. A path starting at 1 can reach 0 only in an odd number of steps, say $2k + 1$. If its first visit at 0 is at time $2k + 1$, this means that it consists of a $2k$-loop at 1 that has the peculiarity of visiting only positive integers (call it a *positive loop*), followed by a one-step transition from 1 to 0. The number of such positive loops of length $2k$ is a well-known classic in combinatorics (see [34], p. 173), the Catalan number $C_k := \frac{1}{k+1}\binom{2k}{k}$. This gives

$$P(T_0 = 2k + 1 | X_0 = 1) = (1 - p)\frac{1}{k+1}\binom{2k}{k}(p(1-p))^k.$$

Summing over k and taking the complement, we obtain

$$P(T_0 \geq 2n + 3 | X_0 = 1) = 1 - (1 - p)\sum_{k=0}^{n}\frac{1}{k+1}\binom{2k}{k}(p(1-p))^k.$$

Letting $n \to \infty$, we obtain

$$P(T_0 = \infty | X_0 = 1) = 1 - (1 - p)F(p(1 - p)),$$

where F is the generating function of the Catalan numbers:

$$f(x) := \sum_{k=0}^{\infty} C_k x^x.$$

Using the ratio test, you may check that the radius of convergence of this series is $\frac{1}{4}$; the exact expression (see [34], p. 178, for a combinatorial proof or Problem 5.2 for a calculus proof) is

$$F(x) = \frac{1 - \sqrt{1 - 4x}}{2x}, \quad 0 < x < \frac{1}{4}. \tag{5.5}$$

Using this expression after some simplification, we obtain $F(p(1-p)) = \frac{1}{p}$ for $p \geq \frac{1}{2}$, and $F(p(1-p)) = \frac{1}{1-p}$ for $p \leq \frac{1}{2}$. In either case we can find the value of $P(T_0 = \infty | X_0 = 1)$, and taking the complement, we obtain the case $k = 1$ of (5.4). For $k \geq 2$ we are going to make use of (1.19). First note that

$$h_{\{0\}}(k) := P(\exists n \geq 0 : X_n = 0 | X_0 = k)$$
$$= P(\exists n \geq 1 : X_n = 0 | X_0 = k) = P(T_0 < \infty | X_0 = k). \quad (5.6)$$

Using (1.19), we obtain

$$\begin{cases} h_{\{0\}}(0) = 1, \\ k \geq 2: \quad h_{\{0\}}(k) = (1-p)h_{\{0\}}(k-1) + ph_{\{0\}}(k+1). \end{cases}$$

Let $u_k := h_{\{0\}}(k)$. Our last equation is a recurrence relation, which we first transform by singling out u_{k+1}:

$$u_{k+1} = \frac{1}{p}u_k - \frac{1-p}{p}u_{k-1}.$$

Subtracting u_k, we get

$$u_k - u_{k+1} = \frac{1-p}{p}(u_{k-1} - u_k) \quad k \geq 1,$$

and thus

$$u_k = 1 - \left[1 + \frac{1-p}{p} + \cdots + \left(\frac{1-p}{p}\right)^{k-1}\right](1 - u_1), \quad k \geq 1.$$

We already know that u_1 is given by (5.4); replacing this in the expression for u_k, we see that $u_k = 1$ for all k if $p \leq \frac{1}{2}$; if $p > \frac{1}{2}$, we may use

$$1 + \frac{1-p}{p} + \cdots + \left(\frac{1-p}{p}\right)^{k-1} = \frac{1 - \left(\frac{1-p}{p}\right)^k}{1 - \frac{1-p}{p}},$$

which gives the second case in (5.4). $\square$

Let us proceed with the determination of stationary distributions. Writing $\pi = \pi p$ in components, you will see that a vector π is invariant if and only if for every $k \geq 0$ it satisfies $\pi_k = \pi_0 r^k$, with $r := \frac{p}{1-p}$. This means that a stationary distribution exists if and only if $r < 1$, i.e., if and only if $p < \frac{1}{2}$. In this case the chain is ergodic, so the convergence result for ergodic chains (Theorem 1.39) applies. The constant π_0 is determined by the fact that π is a probability vector; thus

$$\pi_0(1 + r + \cdots + r^k + \cdots) = 1,$$

i.e., $\pi_0 = 1 - r$. If $p \geq \frac{1}{2}$, there is no stationary distribution, so by Theorem 1.34, the chain is either transient or null recurrent. The previously computed hitting probabilities tell us which case we are in: for $p = \frac{1}{2}$, equation (5.4) implies that 0 is recurrent; thus the chain is null recurrent. For $p > \frac{1}{2}$, it implies that we are in the situation of Proposition 1.35, and the chain is transient.

Summarizing, we have shown the following:

Theorem 5.4. *The reflecting walk on $\mathbb{N}$ is*

$$\begin{cases} \text{positive recurrent for} & p < \frac{1}{2}; \\ \text{null recurrent for} & p = \frac{1}{2}; \\ \text{transient for} & p > \frac{1}{2}; \end{cases}$$

In the positive recurrent case, the stationary distribution is given by

$$\pi_k = \frac{1 - 2p}{1 - p} \left(\frac{p}{1 - p} \right)^k,$$

and it satisfies the local balance relations.

5.1.4 Extinction of a discrete population

A basic model of discrete population dynamics is a variant of the reflecting walk on $\mathbb{N}$, in which the origin is an absorbing point. Let us therefore consider the one-step transition matrix p defined on $\mathbb{N}$ by

$$p(k, k+1) = p_x, \; p(k, k-1) = 1 - p_x := q_x \text{ for } \; k \geq 1,$$

with the absorption condition $p(0,0) = 1$. One important issue from the modeling point of view is that of extinction in finite time. Mathematically, this means that denoting by $(X_n)_{n \in \mathbb{N}}$ the corresponding Markov chain, we are interested in the hitting probabilities $h(k) := P(T_0 < \infty | X_0 = k)$ for all $k \geq 1$. These are computed by the same method as the one we used for the reflecting random walk, the only minor difference being the behavior at the origin. Using (1.19) and proceeding exactly as in the proof of (5.4), we obtain the expression

$$k \geq 1 : h(k) = h(1) - [r_1 + r_1 r_2 + \cdots + r_1 * \cdots * r_{k-1}](1 - h(1)),$$

in which the $r_k := \frac{q_k}{p_k}$ measure the relative importance of deaths versus births. If the series $r_1 + r_1 r_2 + \cdots + r_1 * \cdots * r_{k-1} + \cdots$ diverges, then we must have $h(1) = 1$, and therefore $h(k) = 1$ for all k; this means that whatever the initial data, the population will become extinct in finite time. Note that in the constant case, this happens if and only if the constant $r := \frac{q}{p}$ is greater than or equal to 1.

5.2 Birth–death processes

Birth–death processes provide a generalization of the Poisson process, in which it is assumed that jumps may assume the value -1 as well as $+1$, and no other value. This means that the rate matrix will have nonzero terms on the diagonal and in each row immediately to the left and the right of the diagonal, whence the general form

$$
Q = \begin{pmatrix}
-\lambda_0 & \lambda_0 & & & \\
\mu_1 & -(\lambda_1 + \mu_1) & \lambda_1 & & \\
& \cdot & \cdot & \cdot & \\
& & \cdot & \cdot & \cdot \\
& & \mu_k & -(\lambda_k + \mu_k) & \lambda_k & \cdot \\
& & & \cdot & \cdot \\
& & & & \cdot & \cdot
\end{pmatrix}, \tag{5.7}
$$

in which the λ_k's and μ_k's are nonnegative constants. In order to avoid absorbing states, we shall assume that $\lambda_k + \mu_k \neq 0$ for all k. The coefficients λ_k and μ_k are respectively called the *birth* and *death* coefficients: if X_t is the (integer-valued) number of individuals in a population, then X_t is increased by 1 in case of a birth, and decreased by 1 in case of a death. If $X_t = i$, then the waiting time for a birth is an exponential variable of parameter λ_i, and the waiting time for a death is an exponential variable of parameter μ_i. By Proposition B.5, it follows that the next event (be it a birth or a death) has the distribution $\mathcal{E}(\lambda_i + \mu_i)$, and that this event is a birth with probability $\frac{\lambda_i}{\lambda_i + \mu_i}$, and a death with probability $\frac{\mu_i}{\lambda_i + \mu_i}$. This means that the jump chain is a random walk on $\mathbb{N}$ that jumps by $+1$ with probability $\frac{\lambda_i}{\lambda_i + \mu_i}$ and by -1 with probability $\frac{\mu_i}{\lambda_i + \mu_i}$ (note that this is consistent with the expression for the jump matrix that we obtain if we apply (4.32) to (5.7)).

The process is a *pure birth process* if $\mu_k = 0$ for all k, and the Poisson process is a pure birth process with constant birth rates: $\lambda_k = \lambda$ for all k. Note that we recover the fact that in this case, the jump chain is deterministic: from the value n the only possible transition is to $n+1$. If we write $u_k(t) := P(X_t = k)$, the Fokker–Planck equation written in components is the following:

$$
\begin{cases}
u_0'(t) = -\lambda_0 u_0(t) + \mu_1 u_1(t); \\
u_k'(t) = \lambda_{k-1} u_{k-1}(t) - (\lambda_k + \mu_k) u_k(t) + \mu_{k+1} u_{k+1}(t), \quad k \geq 1.
\end{cases} \tag{5.8}
$$

5.2.1 Stationary distribution

Solving the balance equation $\pi Q = 0$ is equivalent to finding a probability vector π for which the right-hand side of (5.8) vanishes; the equations may be used recursively to express π_1 in terms of π_0, then π_2 in terms of π_1 and π_0 and therefore in terms of π_0, and so on. This yields the candidate $\pi_k = r_k \pi_0$, where

$$
r_k := \frac{\lambda_0 \cdots \lambda_{k-1}}{\mu_1 \cdots \mu_k}, k \geq 1. \tag{5.9}
$$

If all death rates are nonzero, this means that we have found an invariant distribution r (to complete the notation, set $r_0 = 1$); if the sum of its components is finite, we may normalize it to produce a stationary distribution:

$$\pi_0 := \frac{1}{1 + r_1 + \cdots}, \quad \pi_k = r_k \pi_0.$$

In fact, it may be checked that this satisfies the local balance equations: we want to check that

$$Q(k,j)r_k = Q(j,k)r_j$$

for all $k, j \geq 0$ (note that the common factor π_0 plays no role here). For $k = 0$, the only relevant value for j is 1, and the corresponding relation may be checked; for $k > 1$, given the structure of the matrix Q, the relevant values for j are $k - 1$ and $k + 1$, which means that we need to check the equalities (which in fact turn out to be twice the same equality)

$$\mu_k r_k = \lambda_{k-1} r_{k-1}, \quad \lambda_k r_k = \mu_{k+1} r_{k+1}. \tag{5.10}$$

These are immediate consequences of the definition of r, so the local balance equations are indeed satisfied. In the language of population dynamics, (5.10) says that for any size of population k, the loss of individuals through death is exactly balanced by births. Let us summarize:

Theorem 5.5. *Let the coefficients r_k be defined by (5.9), and $r_0 = 1$. Assume that the sum $r_0 + r_1 + \cdots$ is finite. Then the generator Q defined by (5.7) is reversible and has a unique stationary distribution π given by $\pi_k = r_k \pi_0$.*

Let us see what these expressions become in the constant-coefficient case. If $\lambda_k = \lambda$ and $\mu_k = \mu$ for all k, define $p := \frac{\lambda}{\mu}$. If $p < 1$, we obtain

$$\pi_k = (1 - p)p^k, \quad k \geq 0. \tag{5.11}$$

We recognize a shifted geometric distribution of parameter p: if a variable X has probability mass function π, then $X - 1 \sim \mathcal{G}(p)$.

5.2.2 Solution by the Laplace transform method

The system (5.8) having constant coefficients, it lends itself to the use of the Laplace transform; if U_k is the Laplace transform of u_k, we obtain the following recurrence relations:

$$\begin{cases} (\lambda_0 + z)U_0(z) = u_0(0) + \mu_1 U_1(z); \\ (\lambda_k + \mu_k + z)U_k(z) = u_k(0) + \lambda_{k-1}U_{k-1}(z) + \mu_{k+1}U_{k+1}(z), \quad k \geq 1. \end{cases} \tag{5.12}$$

In the pure birth case (meaning $\mu_k = 0$ for all k) these may be solved exactly. The system becomes

$$\begin{cases} (\lambda_0 + z)U_0(z) = u_0(0); \\ (\lambda_k + z)U_k(z) = u_k(0) + \lambda_{k-1}U_{k-1}(z), \quad k \geq 1. \end{cases} \tag{5.13}$$

The solution is obtained from the following elementary lemma, which is easily established by induction:

Lemma 5.6. *Let* $(a_k)_{k \in \mathbb{N}}$, $(b_k)_{k \in \mathbb{N}}$, $(c_k)_{k \in \mathbb{N}}$ *be three given sequences with* $a_k \neq 0$ *for all* k, *and let* $(u_k)_{k \in \mathbb{N}}$ *satisfy the recurrence relations*

$$\begin{cases} a_0 u_0 = c_0; \\ a_k u_k = b_{k-1} u_{k-1} + c_k, \quad k \geq 1. \end{cases} \tag{5.14}$$

Then u_k *is given by*

$$u_k = \frac{c_k}{a_k} + \sum_{j=0}^{k-1} c_j \frac{b_j \cdots b_{k-1}}{a_j \cdots a_k}, \quad k \geq 1.$$

Applying this lemma, we obtain a closed form for U_k:

$$\begin{cases} U_0(z) = \frac{u_0(0)}{\lambda_0 + z}, \\ U_k(z) = \frac{u_k(0)}{\lambda_k + z} + \sum_{j=0}^{k-1} u_j(0) \frac{\lambda_j \cdots \lambda_{k-1}}{(\lambda_j + z) \cdots (\lambda_k + z)}, \quad k \geq 1. \end{cases} \tag{5.15}$$

The next step is to apply the inverse Laplace transform to recover u_k. This can be done easily in two cases: when all birth coefficients are equal and when they are all distinct. Let us begin with constant coefficients: if $\lambda_k = \lambda$ for all k, the expression for U_k becomes

$$U_k(z) = \sum_{j=0}^{k} u_j(0) \frac{\lambda^{k-j}}{(\lambda + z)^{k-j+1}}, \quad k \geq 0.$$

The inverse transform is then given by (A.10). We obtain the following result:

Theorem 5.7 (The Poisson process again). *Assume that* $\mu_k = 0$, $\lambda_k = \lambda$ *for all* k; *then the solution to* (5.8) *is given by*

$$u_k(t) = \sum_{j=0}^{k} u_j(0) \frac{(\lambda t)^{k-j}}{(k-j)!} e^{-\lambda t}, \quad k \geq 0,$$

and the corresponding transition function is the Poisson semigroup:

$$p_t(i, k) = \frac{(\lambda t)^{k-i}}{(k-i)!} e^{-\lambda t}, \, k \geq i.$$

The expression for $p_t(i, k)$ follows from taking $u_j(0) = \delta_{ij}$ in the formula for $u_k(t)$; in problem 5.6 you will see another method of proof for this result.

If we assume that the λ_k's are all distinct, the inverse Laplace transform of the rational fraction $[(\lambda_j + z) \cdots (\lambda_k + z)]^{-1}$ is given by equation (A.9); we obtain the solution for the pure birth process:

Theorem 5.8 (Pure birth processes). *Assume that $\mu_k = 0$ for all k, and that the birth rates λ_k are distinct; then the solution to (5.8) is given by*

$$
\begin{cases}
u_0(t) = u_0(0)e^{-\lambda_0 t}; \\
u_k(t) = u_k(0)e^{-\lambda_k t} + \displaystyle\sum_{j=0}^{k-1} u_j(0)\Big(\prod_{r=j}^{k-1} \lambda_r \Big) \sum_{l=j}^{k} e^{-\lambda_l t} \prod_{m=j, m\neq l}^{k} \frac{1}{\lambda_m - \lambda_l}, \quad k \geq 1.
\end{cases}
$$

$$(5.16)$$

The transition semigroup follows immediately:

Corollary 5.9. *Under the assumptions of Theorem 5.8, the transition semigroup is given by*

$$
p_t(i,k) = \begin{cases}
0 & \text{for} \quad k < i; \\
e^{-\lambda_i t} & \text{for} \quad k = i; \\
\Big(\displaystyle\prod_{r=i}^{k-1} \lambda_r \Big) \sum_{l=i}^{k} e^{-\lambda_l t} \prod_{m=i, m\neq l}^{k} \frac{1}{\lambda_m - \lambda_l} & \text{for} \quad k > i.
\end{cases}
$$

$$(5.17)$$

Proof. We just need to take $u_j(0) = \delta_{ij}$ in the expression for $u_k(t)$; note that this selects the ith term in the sum over j, and this term is present if and only if $k > i$. $\square$

A famous special pure birth process is the Yule process, for which $\lambda_k = k\lambda$, where λ is a positive constant. The above formulas give the following:

Theorem 5.10 (The Yule process). *Assume that $\mu_k = 0$ and $\lambda_k = k\lambda$ for all k; then the solution to (5.8) is given by*

$$
u_k(t) = \sum_{j=0}^{k} u_j(0) \binom{k-1}{j-1} e^{-\lambda j t}(1 - e^{-\lambda t})^{k-j}, \quad k \geq 1,
$$

$$(5.18)$$

and the transition semigroup is given by

$$
p_t(i,k) = \begin{cases}
0 & \text{for} \quad k < i, \\
\binom{k-1}{i-1} e^{-\lambda i t}(1 - e^{-\lambda t})^{k-i} & \text{for} \quad k \geq i.
\end{cases}
$$

$$(5.19)$$

Note that (5.19) has a very simple probabilistic interpretation: conditional on $\{X_0 = i\}$, the variable X_t follows the negative binomial distribution (if needed, see Appendix B for details on the negative binomial distribution) of parameters $e^{-\lambda t}$ and i.

Proof. If $\lambda_k = k\lambda$, the products appearing as coefficients in (5.16) take the following form (to derive the second form, for $j \leq l \leq k$ and $j \leq m \leq k$ locate the positive and negative terms in the product):

$$\prod_{r=j}^{k-1}(r\lambda) = \lambda^{k-j}\frac{(k-1)!}{(j-1)!}, \quad \prod_{m=j,m\neq l}^{k}(m\lambda - l\lambda) = (-1)^{j-l}\lambda^{k-j}(l-j)!(k-l)!.$$

Inserting this result in (5.16), we obtain for $k \geq 1$ that

$$u_k(t) = u_k(0)e^{-\lambda_k t} + \sum_{j=0}^{k-1} u_j(0)\frac{(k-1)!}{(j-1)!}\sum_{l=j}^{k}\frac{e^{-\lambda l t}}{(l-j)!(k-l)!(-1)^{j-l}}.$$

Let us shift the last sum by setting $r = l - j$:

$$u_k(t) = u_k(0)e^{-\lambda_k t} + \sum_{j=0}^{k-1} u_j(0)\frac{(k-1)!}{(j-1)!}\sum_{r=0}^{k-j}\frac{(-1)^r e^{-\lambda(r+j)t}}{r!(k-r-j)!}$$

$$= u_k(0)e^{-\lambda_k t} + \sum_{j=0}^{k-1} u_j(0)\binom{k-1}{j-1}e^{-\lambda j t}\sum_{r=0}^{k-j}\binom{k-j}{r}(-e^{-\lambda t})^r.$$

Equation (5.18) now follows from the binomial formula; the expression (5.19) for the semigroup follows by taking $u_j(0) = \delta_{kj}$, or also as a special case of equation (5.17).   □

5.3 Problems

5.1. Mark $L+1$ points on a circle, and consider a discrete-time random walk that jumps from any point to either one of its two immediate neighbors, clockwise with probability p and counterclockwise with probability $1 - p$.

1. Give the transition matrix.
2. Explain why all column sums of this matrix are 1, and deduce the existence of a stationary distribution.

5.2. In this problem we derive the formula (5.5) for the generating function of the Catalan numbers by elementary calculus; define the function f by $f(x) := \sqrt{1 - 4x}$ for $0 \leq x < \frac{1}{4}$.

1. Show that the Taylor expansion of f is

$$f(x) = 1 - 2x - 2\sum_{k=2}^{\infty}\frac{2^{k-1}}{k!}[3 * 5 * \cdots * (2k - 3)]x^k$$

(in which the expression in brackets is the product of all odd integers up to $2k - 3$).
2. Show that

$$\frac{(2k)!}{k!} = 2^k[3 * 5 * \cdots * (2k - 1)].$$

3. Deduce (5.5).

115

5.3. In the combinatorial proof of Pólya's theorem for $d > 2$, the key point was to use the logarithmic convexity of the gamma function to get a lower bound for $n_1! \cdots n_d!$. Show directly that if $n_1 + \cdots + n_d = n$, then

$$n_1! \cdots n_d! \geq ([\tfrac{n}{d}]!)^d,$$

in which as usual, $[\cdot]$ means integer part. What does this inequality say about the uniform multinomial distribution?

5.4. For the symmetric random walk on $\mathbb{Z}$, how likely is a return to be the first one? Given that after $2n$ steps the walk has returned to its original position, compute the conditional probability that this is the first return.

5.5. The goal of this problem is to give a taste of the so-called *generating function method*. For the symmetric random walk on $\mathbb{Z}$, let p_n be the probability of returning to 0 after n steps given that the original position is 0, and let q_n be the probability of doing this without ever hitting 0 before time n. Let P and Q be their generating functions:

$$P(z) := \sum_{k=0}^{\infty} p_k z^k, \quad Q(z) := \sum_{k=0}^{\infty} q_k z^k,$$

for every complex number z such that the series converges. Show that

$$q_n = \sum_{k=0}^{n} p_k q_{n-k},$$

and deduce the relation $P(z)Q(z) = Q(z) - 1$.

5.6. We proved Theorem 5.7 by making use of the Laplace transform in order to solve the Fokker–Planck equation; an alternative method consists in using the generating function of the sequence $(u_k(t))_{k \in \mathbb{N}}$ (recall that $u_k(t) := p_t(0, k)$), which as in previous problem is defined as

$$\phi(z, t) := \sum_{k=0}^{\infty} z^k u_k(t).$$

In more probabilistic terms, $\phi(\cdot, t)$ is the probability generating function of X_t conditional on $X_0 = 0$.

1. In the case of constant coefficients $\lambda_k = \lambda$, $\mu_k = \mu$, show that ϕ satisfies the differential equation

$$\frac{\partial \phi}{\partial t} = \frac{\lambda z^2 - (\lambda + \mu)z + \mu}{z} \phi(z, t) + \mu(1 - \frac{1}{z})u_0(t).$$

2. If, moreover, $\mu = 0$, use this equation to recover the Poisson semigroup.

5.7. Consider a finite birth–death chain defined on the set of integers $\{0, \ldots, N\}$ by the rate matrix A:

$$\begin{cases} A(n, n+1) = \lambda_n & \text{for} \quad 0 \leq n < N, \\ A(n, n-1) = \mu_n & \text{for} \quad 0 < n \leq N. \end{cases}$$

Find a stationary distribution, and show that the local balance equations are satisfied.

Part II

Entropy and applications

6

Prelude: a user's guide to convexity

Summary. This chapter could be called a primer on convexity for probabilists, since it brings together some of the theory of convex functions with a view toward applications to information theory. In particular, a certain number of classical convexity inequalities (such as Jensen's and Young's inequalities) are presented in a unified framework thanks to the use of Bregman's divergences.

6.1 Why are convex sets and convex maps so important in probability?

Let us begin with an informal discussion on the relevance of convexity for probability and information theory. Some of the most basic and most fundamental relations you have learned in discrete probability are the formula for computing the expectation of a function of a random variable and the total probability theorem:

$$E(\phi(X)) = \sum_{x} \phi(x)P(X = x), \qquad (6.1)$$

$$P(A) = \sum_{i} P(A|B_i)P(B_i). \qquad (6.2)$$

If we forget for a moment about the probabilistic meaning of these relations and focus on the algebra, we see that both relations consist in taking a linear combination of some arbitrary numbers (the $\phi(x)$ in the first case and the $P(A|B_i)$ in the second) with some nonnegative coefficients that add up to 1. This is what is called a *convex combination*:

Definition 6.1. *Let $x_1, x_2, \ldots, x_r$ be a finite collection of points in $\mathbb{R}^n$; a convex combination of these points is a point of the form $\lambda_1 x_1 + \cdots + \lambda_r x_r$, where all λ_i are between 0 and 1, and $\sum_{i=1}^{r} \lambda_i = 1$.*

Convex combinations may be thought of probabilistically or geometrically: in probabilistic language, convex combinations of the points $x_1, x_2, \ldots, x_r$

© Springer International Publishing AG, part of Springer Nature 2018

J.-F. Collet, *Discrete Stochastic Processes and Applications,*

Universitext, https://doi.org/10.1007/978-3-319-74018-8_6

represent all possible values for $E(X)$, for X an $\mathbb{R}^n$-valued random variable assuming the values $x_1, x_2, \ldots, x_r$. In pen and paper language, for $n = 2$ take a set of points $x_1, \ldots, x_r$ and draw the largest possible polygon each edge of which runs from some x_i to another; the set of convex combinations of the x_i is the inside of this polygon (edges included of course). This set C, also known as the *convex hull* of the set $\{x_1, \ldots, x_r\}$, obviously has the property that for every two points $x_1, x_2 \in C$, the segment $[x_1, x_2]$ lies entirely in C.

Why is this relevant to probability? The key point here is that a large chunk of probability theory is about estimating (which means either bounding from above or below) probabilities or expectations when these cannot be computed exactly, and convex combinations give us inequalities. Here is an example: from (6.1) and (6.2) you can immediately show the inequalities

$$\min_x \phi(x) \leq E(\phi(X)) \leq \max_x \phi(x), \quad \min_i P(A|B_i) \leq P(A) \leq \max_i P(A|B_i).$$

The first two inequalities tell us that taking expectations is an averaging process, so the result lies between the extrema of ϕ, and in the last two inequalities the lower bound is attained for the "least favorable B_i for A" (there will be at least one), while the upper bound is attained for the "most favorable B_i for A." A very slightly less elementary example is given by Markov matrices: if p is a probability vector and P is a Markov matrix, then from the definition of the product pP we immediately get the bounds

$$\min_x p(x) \leq \min_x (pP)(x) \leq \max_x (pP)(x) \leq \max_x p(x).$$

In other words, multiplication of a probability vector by a Markov matrix has a regularizing (here the word regularization is used in a loose sense by analogy with functions) effect: minima go up; maxima go down. The vector pP is in some sense "flatter," "less random" than p (these vague statements will be made precise when we learn later about entropy). If we think of the Chapman–Kolmogorov equation now, we see that more precise bounds could tell us something about the asymptotic behavior of Markov chains, which is why it might be worth taking a detour by convex analysis.

The theory of convex functions and convex maps is in fact a whole branch of mathematics; our goal here is to collect a few basic facts about convex functions that are used in the text, focusing on the general ideas rather than trying to give optimal results and detailed proofs; for a more thorough treatment see, for instance, [13, 31].

Let us begin with convex sets:

Definition 6.2. *A set $C \subset \mathbb{R}^n$ is said to be convex if*

$$\forall \lambda \in [0,1], \forall x_1, x_2 \in C : \lambda x_1 + (1 - \lambda)x_2 \in C.$$

You should have no difficulty proving that this is equivalent to the same condition for general convex combinations (if you find it difficult, look at the proof of Theorem 6.8 below for inspiration):

Theorem 6.3. *A set $C \in \mathbb{R}^n$ is convex if and only if it is closed under the operation of convex combinations, that is, for every finite collection of points $x_1, \ldots, x_r$ of points of C, all convex combinations of the x_i also belong to C.*

So now that we know about convex sets, what about convex maps?

Definition 6.4. *A map $f : \mathbb{R}^n \to \mathbb{R} \cup \{+\infty\}$ is said to be convex if its domain $D(f)$ is convex and*

$$\forall \lambda \in [0,1], \forall x_1, x_2 \in D(f) : f(\lambda x_1 + (1-\lambda)x_2) \leq \lambda f(x_1) + (1-\lambda)f(x_2). \quad (6.3)$$

The map f is said to be concave if $-f$ is convex.

In words: take two points on the graph and draw the segment between these two. The condition says that this segment lies entirely above the curve (see Problem 6.2 below for a precise statement). At this point the reader is urged to graph the functions $|z|, z^2$, and z^4, and convince themselves by a picture that these are convex functions. This is heuristics, not mathematics. The mathematical blanks are filled in by actually proving (6.3), which for at least the first two functions should not be a problem. While we're at it, also graph the function $z \ln z$, and convince yourself by a picture that it is convex. More on this later.

Just as for sets, it is easy to check that the two-point condition above implies the analogous r−point condition for every r. This fact is known as *Jensen's inequality:*

Theorem 6.5. *Let $f : \mathbb{R}^n \to \mathbb{R}$ be convex. Then for every choice of the weights $\lambda_1, \ldots, \lambda_k$ satisfying $0 \leq \lambda_j \leq 1$ and $\sum_{j=1}^k \lambda_j = 1$, we have*

$$\forall x_1, \ldots, x_k \in D(f) : \quad f\left(\sum_{j=1}^k \lambda_j x_j\right) \leq \sum_{j=1}^k \lambda_j f(x_j). \quad (6.4)$$

If we think of a random variable X assuming the value x_i with probability λ_i, we see that Jensen's inequality translates as follows:

Theorem 6.6. *For a discrete random variable X and convex function ϕ defined on the range of X, we have the inequality*

$$\phi(E(X)) \leq E(\phi(X)). \quad (6.5)$$

(If the function ϕ is differentiable, then we will see in the section on Bregman divergences a very quick proof of this result.) Note in particular that for $\phi(z) = z^2$ this gives the well-known property $(E(X))^2 \leq E(X^2)$, which is tantamount to saying that $var(X) \geq 0$.

As mentioned above, the main two prototypes of convex functions in one dimension are of course the absolute value and the square function. Leaving aside differentiability issues for the moment, the main difference between these two is that inequality (6.3) happens to be an equality quite often (for instance for every λ as soon as x_1 and x_2 have the same sign) for the absolute value, but never for the square unless $\lambda x_1 + (1 - \lambda)x_2$ happens to be either x_1 or x_2. In other words, the graph of the absolute value function contains segments, whereas the graph of the square function does not. Let us formalize the way to discriminate these two types of situations:

Definition 6.7. *A convex function f is said to be strictly convex if the equality*

$$f(\lambda x_1 + (1 - \lambda)x_2) = \lambda f(x_1) + (1 - \lambda)f(x_2)$$

implies $x_1 = x_2$, $\lambda = 1$, or $\lambda = 0$.

The next result (quite similar in form to Theorem 6.3) shows that this also characterizes the case of equality in Jensen's inequality for an arbitrary number of points:

Theorem 6.8. *Let f be strictly convex. Then for every integer $r \geq 2$, if a convex combination $\lambda_1 x_1 + \cdots + \lambda_r x_r$ of points in $D(f)$ satisfies*

$$f\left(\sum_{j=1}^{r} \lambda_j x_j\right) = \sum_{j=1}^{r} \lambda_j f(x_j), \tag{6.6}$$

then either these points are all equal or one of the weights λ_i is equal to 1.

Proof. Of course there is nothing to show for $r = 2$, so we are all set to prove this by induction. Assuming that the property has been established for all integers up to $r - 1$, we may assume that $\lambda_1 \neq 1$ (otherwise, we are done), and in fact that all weights are strictly positive (otherwise, we are dealing with fewer than r points, and we are done). Let us isolate x_1 in the sum and set $\mu_i := \frac{\lambda_i}{1-\lambda_1}$ for $i \geq 2$; note that $\mu_2 + \cdots + \mu_r = 1$. Using Jensen's inequality first with $r - 1$ points (with weights μ_i) and then with 2 points, we get

$$\sum_{i=1}^{r} \lambda_i f(x_i) = (1 - \lambda_1) \sum_{i=2}^{r} \mu_i f(x_i) + \lambda_1 f(x_1)$$

$$\geq (1 - \lambda_1) f\left(\sum_{i=2}^{r} \mu_i x_i\right) + \lambda_1 f(x_1)$$

$$\geq f\left((1 - \lambda_1) \sum_{i=2}^{r} \mu_i x_i + \lambda_1 x_1\right)$$

$$= f\left(\sum_{i=1}^{r} \lambda_i x_i\right).$$

Since (6.6) holds, both inequalities above have to be equalities; from the first one we obtain

$$\sum_{i=2}^{r} \mu_i f(x_i) = f\left(\sum_{i=2}^{r} \mu_i x_i\right).$$

Using now the property for $r - 1$, we must have either $\mu_i = 1$ for some $i \geq 2$
or

$$x_2 = x_3 = \cdots = x_r. \tag{6.7}$$

If $\mu_i = 1$, then all other μ_k are equal to zero for $k \neq i$, and $\lambda_k = 0$, which is
impossible, so in fact, (6.7) must be satisfied; returning to (6.6), we now have

$$f(\lambda_1 x_1 + (1 - \lambda_1)x_r) = \lambda_1 f(x_1) + (1 - \lambda_1)f(x_r),$$

and from the two-point property we see that $x_1 = x_r$, so that all points are
equal.   □

The translation into probabilistic terms is the following:

Theorem 6.9 (case of equality in Jensen's inequality). *Let X be a discrete random variable, and let ϕ be a strictly convex function defined on the
range of X. If $E(\phi(X)) = \phi(E(X))$, then X is a constant variable.*

How to tell whether a given function is convex? The definition, although
geometrically quite intuitive, is not very convenient to use. It involves infinitely
many inequalities that might be hard to prove directly, so we would like to
have a way of checking convexity more directly. There are basically two situations in which this can be done, namely when f is obtained from a set of
convex functions by operations that preserve convexity, and when the second
derivative of f is a positive semidefinite matrix. Let us begin with convexity-
preserving operations. The following results are fairly easy to prove, so to
maintain consistency with our user-oriented approach to convexity we will
just state them:

Theorem 6.10 (Operations that preserve convexity). *Convexity is preserved by linear combinations with nonnegative coefficients, composition with
increasing convex functions, and taking the supremum. More precisely:*

- *If $f_1, \ldots, f_r$ is a finite family of convex functions all defined on a set
  $D \subset \mathbb{R}^n$ and the numbers $\lambda_1, \ldots, \lambda_r$ are all nonnegative, then the map
  $f := \lambda_1 f_1 + \cdots + \lambda_r f_r$ is convex on D; if all the f_i are strictly convex and
  all the λ_i are positive, then f is strictly convex.*
- *If f is a convex map and $g : \mathbb{R} \to \mathbb{R}$ is an increasing convex map defined
  on the range of f, then the map $h : D(f) \to \mathbb{R}$ defined by $h(x) = g(f(x))$ is
  convex; if g is strictly increasing and f is strictly convex, then h is strictly
  convex.*
- *If $\mathcal{F}$ is a family of convex maps all defined on a set $D \subset \mathbb{R}^n$, then the map
  f that is defined pointwise on D by*

$$f(x) := \sup_{g \in \mathcal{F}} g(x)$$

is convex on D.

For the case of the supremum, note that we do not have a "strictly convex" refinement of the statement as we did for the first two cases; this should not surprise you, since you know that strict inequalities do not remain strict when a supremum is taken. Also note that we do not require the family $\mathcal{F}$ to be finite (in fact, it does not even need to be countable). Perhaps the most common application of the second point, composition with an increasing function, is the case $g(z) = e^z$. In this case, it may be reformulated as follows: if $\ln f$ is convex, then f is convex. This situation is so frequent that it has a name:

Definition 6.11. *A function $f : \mathbb{R}^n \to \mathbb{R}$ with positive values is said to be log-convex if the function $\ln f$ is convex.*

So all log-convex functions are convex; the converse is grossly false! just take $f(x) := x^2$ on $\mathbb{R}$.

As mentioned above, another way to determine whether a function is convex is to look at its second derivative (provided it has one, of course):

Theorem 6.12. *Let $f : \mathbb{R}^n \to \mathbb{R}$ be twice differentiable everywhere in its domain; then*

- *The map f is convex if and only if for every $x \in D(f)$, the Hessian $\nabla^2 f(x)$ is a positive semidefinite matrix.*
- *If for every $x \in D(f)$, the Hessian $\nabla^2 f(x)$ is a positive definite matrix, then f is strictly convex.*

The proof (as well as more general and sharper statements) may be found in any textbook on convex analysis; I just wish to draw your attention to the fact that the second point above is only a sufficient condition: just think of the function x^4 on $\mathbb{R}$. What does this say about tangents to the graph? In one dimension, the convexity condition for a C^2 function is simply $f'' \geq 0$, which means that the slope $f'(x)$ of the tangent line is an increasing function of x. Moreover, if we use the second-order Taylor formula, we immediately see that the graph must lie above the tangent everywhere. In fact, this can be shown even if the map f is differentiable only once. Here is the precise statement:

Theorem 6.13. *Assume that $f : \mathbb{R}^n \to \mathbb{R}$ is C^1 on $D(f)$, and that $D(f)$ is convex. Then f is convex if and only if*

$$\forall x, y \in D(f) : f(x) \geq f(y) + \nabla f(y)(x - y),$$

and f is strictly convex if and only if

$$\forall x, y \in D(f), x \neq y : f(x) > f(y) + \nabla f(y)(x - y).$$

If the function is C^2 the proof is immediate using Taylor's formula of order 2; if it is only C^1, we omit the proof; for details consult, for example, [13, 31].

6.2 The Legendre transform

Let us begin with some heuristics in order to provide a geometric motivation for the definition of the Legendre transform.

In one dimension we just saw that for a differentiable convex function, the slope $f'(x)$ varies monotonically, so that roughly speaking, to any prescribed slope (in some appropriate interval) there should correspond a unique point on the graph. In a way, we adopt here a point of view that is dual (whatever this exactly means) to the classical one when one computes the slope of the tangent to the graph at a given point x. We learn in calculus that this is well defined if things go well near x, which precisely means if f is differentiable at x, and after some reasoning we come to the conclusion that the slope of the tangent is $f'(x)$. Of course the slope is all we need, since we already know that the line we are looking for goes through the point $(x, f(x))$. Let us now reverse this point of view and decide that we want to geometrically represent the function f not by its graph but by the set of its tangent lines. We know (since f' is monotone) that all slopes in some interval are possible, and so it just remains to say for each admissible slope what is the corresponding line. The Legendre transform appears when one decides to define this line by its y-intercept. The corresponding formulas are very easy to obtain. Let p be a fixed number. The point on the graph where the tangent has slope p has for first coordinate the unique number x_0 defined by $f'(x_0) = p$; the Cartesian equation of the tangent line is now $y = f(x_0) + p(x - x_0)$, so that its y-intercept is $f(x_0) - px_0$. This quantity is uniquely determined in terms of p, so we may call it $g(p)$. To summarize this two-step construction, for any given p, the quantity $g(p)$ is computed by solving the system

$$f'(x) = p, \quad g(p) = f(x) - px.$$

As we will see below, this would give a concave function; since we prefer to deal with convex functions, we consider the negative of g, which we call f^*; therefore, for suitable p the quantity $f^*(p)$ is uniquely defined by

$$f'(x) = p, \quad f^*(p) + f(x) = px. \tag{6.8}$$

This definition makes sense only if p is in the range of f'; It is immediate to check that in this case, this last system is equivalent to the maximization problem

$$f^*(p) = \sup_x (px - f(x)),$$

and that the supremum is attained (precisely as the unique x satisfying $f'(x) = p$). In the general case, where the supremum may not be attained or may be infinite, we will retain this as the correct definition of the Legendre transform:

Definition 6.14. *Let $f : \mathbb{R}^n \to \mathbb{R}$ be a convex map; the Legendre transform of f is the map f^* defined by*

$$f^*(p) := \sup_x (px - f(x)).$$

127

The informal discussion that we just used to motivate the introduction of the Legendre transform can now be turned into a recipe for its computation:

Theorem 6.15. *Let $f : \mathbb{R}^n \to \mathbb{R}$ be a C^1 convex map. Then for all p in the range of ∇f, the system 6.8 uniquely defines the pair $(x, f^*(p))$.*

In this case, x and p are said to be *conjugate variables*, or the pair (x, p) is said to be a *conjugate pair*. At this point we urge you to check by a direct computation that the Legendre transform of $\frac{1}{2}x^2$ is $\frac{1}{2}p^2$, and that the Legendre transform of e^x is $p(\ln p - 1)$. One thing Legendre transforms are good for is producing inequalities; from the previous theorem and (6.8) we immediately get the following:

Theorem 6.16 (Young's inequality). *Let $f : \mathbb{R}^n \to \mathbb{R}$ be a C^1 convex map. For all p in the range of ∇f and all x in $D(f)$ we have*

$$f^*(p) + f(x) \geq px.$$

For instance, with $f(x) = \frac{1}{2}x^2$ we get the trivial inequality

$$\frac{1}{2}x^2 + \frac{1}{2}p^2 \geq px,$$

which could give the impression that this Legendre machinery is much ado about nothing; see Problem 6.11 for something more exciting.

The two main properties of the Legendre transform are the following:

Theorem 6.17. *Let $f : \mathbb{R}^n \to \infty$ be a convex map. Then f^* also is convex, and $f^{**} = f$.*

The interested reader may consult [31] for a proof. Let us give a pseudo-proof in the case $n = 1$ for a twice differentiable map and assuming that the second derivative f'' never vanishes. If we take the derivative of both relations in (6.8) with respect to x, we obtain

$$\frac{dp}{dx} = f''(x), \quad f^{*'}(p)f''(x) + p = p + f''(x)x,$$

so that

$$f^{*'}(p) = x, f^*(p) + f(x) = px,$$

which shows precisely that $f^{**}(p) = x$. Differentiating the first relation one more time with respect to x, we obtain

$$f^{*''}(p)f''(x) = 1,$$

which shows that f^* is convex. $\square$

6.3 Bregman divergences

Convex functions may be used to evaluate how close two points in $\mathbb{R}^n$ are, using a generalization of the familiar p-norms. Let us begin with a general definition:

Definition 6.18. *Let $\Phi : \mathbb{R}^n \to \mathbb{R}$ be a C^1 strictly convex function. The Bregman divergence associated with Φ is the real-valued map B_Φ defined on $D(\Phi) \times D(\Phi)$ by*

$$\forall p, q \in D(\Phi) : B_\Phi(p, q) := \Phi(p) - \Phi(q) - \nabla\Phi(q)(p - q).$$

In other words, for p close to q, the quantity $B_\Phi(p, q)$ is the remainder in Taylor's formula at point q after the first-order term. Here is a picture to visualize the meaning of B_Φ in the scalar case:

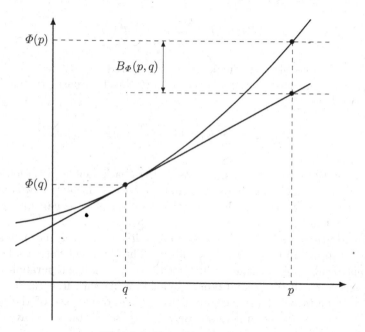

Fig. 6.1. Bregman divergence

Of course, the quantity $B_\Phi(p, q)$ could be defined even for a nonconvex Φ, but the key point here is that since Φ is strictly convex, B_Φ is always nonnegative, and (in view of Theorem 6.13) it separates points:

$$\forall p, q \in \mathbb{R}^n : B_\Phi(p, q) = 0 \Rightarrow p = q. \tag{6.9}$$

In what follows we will deal only with the case in which Φ has the so-called *separated form*, i.e., is of the form

$$\Phi(p) = \sum_x \phi(p(x))$$

for some strictly convex function ϕ, so that B_Φ becomes

$$B_\Phi(p, q) = \sum_x [\phi(p(x)) - \phi(q(x)) - \phi'(q(x))(p(x) - q(x))].$$

By a slight abuse of notation we shall write D_ϕ instead of D_Φ (we leave it to you to check that if ϕ is strictly convex, then Φ also is). Bregman divergences have recently been increasingly used in many fields where probability distributions need to be compared (such as image processing, for instance), either in general form or for some particular choice of ϕ. Let us just mention the two most classical cases:

- $\phi(z) = z^2$: in this case we get

$$B_\Phi(p, q) = \sum_x (p(x) - q(x))^2,$$

  which is the well-known Euclidian norm squared.
- $\phi(z) = z \log z$, where log denotes any logarithm to a base greater than 1: now we have

$$B_\Phi(p, q) = \sum_x [p(x) \log \frac{p(x)}{q(x)} + q(x) - p(x)]. \tag{6.10}$$

For probability vectors this is known as the Kullback–Leibler divergence, and it plays a very important role in information theory; in fact, it was used in information theory prior to the general formalism of Bregman divergences. More on that in the chapter on information theory.

What do these two examples have to do with each other and why should they be wrapped up in a general formalism? The key point is that all convex functions locally look the same, which means they look like a parabola. To be more precise, recall the second-order Taylor formula with integral remainder (calculus experts will have recognized the second-order case of the general Taylor formula with integral remainder, but those of you who haven't met this before may just check it by integration by parts):

$$\phi(y) = \phi(x) + (y - x)\phi'(x) + \int_x^y \phi''(s)(y - s)\, ds. \tag{6.11}$$

Transforming the integral remainder by an affine change of variable, one immediately gets the relation

$$B_\phi(p, q) = (p - q)^2 \int_0^1 \phi''(q + t(p - q))(1 - t)\, dt,$$

which to use nonmathematical language tells us that if ϕ'' does not change too much or if p and q are very close, then $B_\phi(p, q)$ is close to their squared Euclidian distance.

One nice algebraic property of the Bregman divergence is its behavior with respect to the Legendre transform:

Theorem 6.19. *Let $f : \mathbb{R}^n \to \mathbb{R}$ be a C^1 convex function. Then for all p_1, p_2 in the range of ∇f we have*

$$B_f(x_1, x_2) = B_{f^*}(p_2, p_1), \tag{6.12}$$

where x_1 and x_2 are respectively conjugate to p_1 and p_2.

Proof. Using (6.8), we have for $i = 1, 2$,

$$f^*(p_i) = x_i \nabla f(x_i) - f(x_i), \quad \nabla f^*(p_i) = x_i.$$

The equality follows on inserting these in the definition of $B_{f^*}(p_2, p_1)$. $\square$

Bregman divergences provide a nice synthetic way to recast a large family of convexity inequalities; more precisely, for both Young's and Jensen's inequalities the gap between the quantity of interest and its upper bound is in fact a Bregman divergence. Let us begin with Young's inequality:

Theorem 6.20. *Let $f : \mathbb{R} \to \infty$ be a C^1 map. Then for all x, q we have*

$$f^*(q) + f(x) - qx = B_{f^*}(q, p) = B_f(x, y), \tag{6.13}$$

where y is conjugate to q, and p is conjugate to x.

Proof. In view of (6.12), we have only to prove the first equality; this is a straightforward computation:

$$B_{f^*}(q, p) = f^*(q) - f^*(p) - \nabla f^*(p)(q - p)$$
$$= f^*(q) - f^*(p) - x(q - p) = f^*(q) + f(x) - qx. \tag{6.14}$$

$\square$

Moving on to Jensen's inequality, we have the following statement:

Theorem 6.21. *For every random variable X and strictly convex differentiable convex function ϕ, we have*

$$E(\phi(X)) - \phi(E(X)) = E(B_\phi(X, \mu)). \tag{6.15}$$

Proof. Calling respectively p and μ the probability vector and expectation of X, we use the usual probability-vector trick to combine the two sums and then rewrite the ϕ-difference as a Bregman divergence:

$$E(\phi(X)) - \phi(E(X)) = \sum_x \phi(x)p(x) - \phi(\sum_y yp(y))$$

$$= \sum_x [\phi(x) - \phi(\sum_y yp(y))]p(x)$$

$$= \sum_x [B_\phi(x, \mu) + \phi'(\mu)(x - \mu)]p(x)$$

$$= \sum_x B_\phi(x, \mu)p(x).$$

□

This relation gives us an immediate proof of Jensen's inequality. Indeed, in view of (6.9), the right-hand side in the last equality is nonnegative, and it vanishes if and only if X is a constant variable. Such a useful quantity deserves a name:

Definition 6.22. *For a random variable X, the Bregman information of X (relative to ϕ) is the quantity*

$$I_\phi(X) := E(B_\phi(X, \mu)).$$

If $\phi(z) = z^2$, we see that the Bregman information is simply the variance of X; you may have seen in elementary probability (or you may check it just now as an exercise if that is not the case) that the quantity $E((X-c)^2)$ is minimal for $c = \mu$, in which case it is equal to the variance. We can now generalize it to arbitrary ϕ:

Theorem 6.23. *For every constant c the following relation holds:*

$$E(B_\phi(X, c)) - E(B_\phi(X, E(X))) = B_\phi(E(X), c).$$

As a consequence, the quantity $E(B_\phi(X, c))$ is minimal for $c = E(X)$, in which case it is equal to $I_\phi(X)$.

The proof is an immediate adaptation of the proof of (6.15), so we leave it to you as an exercise.

6.4 Problems

6.1. Let $f : \mathbb{R}^n \to \mathbb{R}$ be twice differentiable; using Taylor's formula for functions of one variable (6.11), prove Taylor's formula with integral remainder in many variables:

$$\forall x, y \in \mathbb{R}^n : \quad f(y) - f(x) - \nabla f(x)(y - x)$$

$$= \int_0^1 \nabla^2 f(x + t(y - x))(y - x) \cdot (y - x)(1 - t) \, dt. \tag{6.16}$$

6.2. You all know that the graph of a function $f : \mathbb{R}^n \to \mathbb{R}$ is the set of points of $\mathbb{R}^{n+1}$ of the form $(x, f(x))$, where x runs over the domain of definition $D(f)$ of f, but did you know that the epigraph $e(f)$ of f is the part of $\mathbb{R}^{n+1}$ located above the graph:

$$e(f) := \{(x, y), x \in D(f), y \in \mathbb{R}, \ y \geq f(x)\}.$$

Assuming that the domain $D(f)$ is convex, show that the map f is convex if and only if its epigraph is convex.

6.3. Consider the so-called LogSumExp map f defined on $\mathbb{R}^n$ by

$$f(x) := \ln \sum_{k=1}^n e^{x_k}.$$

The goal is to show that f is convex and to see what inequalities can be deduced from this.

1. Show that the second derivative of f is given by

$$\nabla^2 f(x) = \frac{A}{T^2},$$

where

$$A(i, j) = \delta_{i,j} t_i T - t_i t_j,$$

$$t_i := e^{x_i}, \quad T := \sum_k t_k.$$

2. Check that for every vector $p \in \mathbb{R}^n$,

$$Ap \cdot p = \frac{1}{2} \sum_i \sum_j t_i t_j (p(i) - p(j))^2,$$

and conclude (hint: use Theorem 6.12).
3. Deduce that for every two vectors $p, q \in \mathbb{R}^n$ with all q_i nonzero, the so-called *logsum inequality* holds:

$$\sum_i p(i) \ln \frac{p(i)}{q(i)} \geq \left(\sum_i p(i)\right) \ln \frac{\sum_k p(k)}{\sum_k q(k)}.$$

6.4. How good is the lower bound of last problem? In the section on Bregman divergences we computed $B_\phi(p, q)$ for the function $\phi(z) := z \log z$. Use (6.10) to prove that

$$\sum_i p(i) \ln \frac{p(i)}{q(i)} \geq \sum_k p(k) - \sum_k q(k).$$

How does this lower bound compare to the one you obtained in the previous exercise?

6.5. Interestingly enough, the computations in this exercise are quite similar to what you did in Problem 6.3, but this time they will imply concavity. Consider the map f defined on $[0, \infty)^n$ by

$$f(x) := \left(\prod_{k=1}^{n} x_k\right)^{\frac{1}{n}}$$

(f(x) is called the *geometric mean* of the numbers $x_1, \ldots, x_n$). Show that $\nabla^2 f(x) = f(x)A$, where A is a matrix (to be determined) such that for all $p \in \mathbb{R}^n$,

$$Ap \cdot p = -\frac{1}{2} \sum_i \sum_j \left(\frac{p(j)}{x_j} - \frac{p(i)}{x_i}\right)^2, \tag{6.17}$$

and conclude (hint: use Theorem 6.12).

6.6. A *monomial* is a function defined on $[0, \infty)^n$ of the form

$$f(x) := \prod_{k=1}^{n} x_k^{\alpha_k},$$

where $\alpha_1, \ldots, \alpha_k$ are fixed real numbers (note that some texts put a multiplicative constant in front of $f(x)$, but we prefer to omit it here). Given a monomial f, let us define a new function g on $\mathbb{R}^n$ by setting

$$g(t) = f(e^{t_1}, \ldots, e^{t_n}).$$

Show that g is a convex function. This is what is meant when people casually say "f is a convex function of $\ln x$." As a function of x, though, f may or may not be convex (just think of the case $n = 1$); the goal of the next two problems is to investigate some particular cases.

6.7. In this problem we look at the case in which all exponents of a monomial are negative.

1. As a warmup, compute the second derivative of the map f defined on $\mathbb{R}^2$ by $f(x) := \frac{1}{x_1 x_2}$. Compute its trace and determinant, and conclude that f is convex.
2. Consider now the general case of a monomial f with all exponents negative; use a similar method to show that f is log-convex.

6.8. Now the harder case in which all exponents are positive.

1. Compute the second derivative $\nabla^2 f(x)$ and write it in the form $\nabla^2 f(x) = f(x)A(x)$, where $A(x)$ is a matrix to be specified.
2. Show that the determinant of $A(x)$ is given by

$$\det A(x) = \left(\prod_{i=1}^{n} \frac{\alpha_i}{x_i^2}\right) \det C,$$

where C is the matrix defined by

$$C(i,j) := \alpha_j - \delta_{i,j}.$$

We now want to compute the quantity $D_n(\alpha_1, \ldots, \alpha_n) := \det C$.

3. Compute D_2, and show that

$$D_n(\alpha_1, \ldots, \alpha_n) = -D_{n-1}(\alpha_1, \ldots, \alpha_{n-1}) + (-1)^n \alpha_n.$$

4. Use this last relation to show that

$$D_n(\alpha_1, \ldots, \alpha_n) = (-1)^n (1 - \sum_{i=1}^n \alpha_i).$$

5. Deduce the values of all leading principal minor determinants of C and conclude that if $\sum_{i=1}^n \alpha_i < 1$, then $-f$ is convex (recall that a matrix is positive definite if and only if its leading principal minor determinants are all positive).

6.9. The expression obtained in the last exercise for the determinant of $\nabla^2 f(x)$ (and all its leading principal minors) is valid whatever the exponents α_i are, but for positive α_i, it gave us a conclusion only if $\sum_{i=1}^n \alpha_i < 1$. In the case $\sum_{i=1}^n \alpha_i = 1$ (which includes the case of the geometric mean), the determinant vanishes, and we cannot conclude as above; in this case, show that for all $p \in \mathbb{R}^n$ we have the relation

$$\nabla^2 f(x) p \cdot p = -\frac{1}{2} f(x) \sum_i \sum_j \alpha_i \alpha_j \left(\frac{p(i)}{x_i} - \frac{p(j)}{x_j}\right)^2 \qquad (6.18)$$

(which generalizes (6.17)), and conclude (hint: use Theorem 6.12).

6.10. The last two problems show that if $\alpha_i > 0$ for all i and $\sum_{i=1}^n \alpha_i \leq 1$, the function

$$f(x) := -\prod_{k=1}^n x_k^{\alpha_k}$$

is convex.

1. Use this fact to show that for all $t \in \mathbb{R}^n$ with positive components,

$$\prod_{k=1}^n t_k^{\alpha_k} \leq \sum_{k=1}^n \alpha_k t_k. \qquad (6.19)$$

2. In the case $\sum_{i=1}^n \alpha_i = 1$, compute $B_f(t, s)$, where t is any vector and s is a vector all of whose components are equal to 1, and conclude that equality is attained in (6.19) if and only if all the t_i are equal to 1.

Remark: If all the α_k are equal to $\frac{1}{n}$, then (6.19) says that the geometric mean is less than the arithmetic mean, a fact that is called the *inequality of arithmetic and geometric means*, or the *AM–GM inequality*. Well done! You have just generalized it to arbitrary $\alpha_k > 0$.

6.11. We are going to show that Young's inequality implies Hölder's inequality. Let p be a real number strictly greater than 1, and let f be the function defined on $\mathbb{R}^n$ by

$$f(x) := \frac{1}{p} \sum_{k=1}^{n} |x_k|^p.$$

1. Show that the Legendre transform of f is given by

$$f^*(y) = \frac{1}{q} \sum_{k=1}^{n} |y_k|^q,$$

where q, the so-called *conjugate exponent of p*, is the unique number such that

$$\frac{1}{p} + \frac{1}{q} = 1.$$

2. Using Young's inequality, show that

$$\forall p > 1, \forall x, y \in \mathbb{R}^n : \quad \sum_{k=1}^{n} |x_k y_k| \leq \left(\sum_{k=1}^{n} |x_k|^p \right)^{\frac{1}{p}} \left(\sum_{k=1}^{n} |y_k|^q \right)^{\frac{1}{q}}.$$

3. Deduce the following:

$$\forall \lambda \in [0,1], \forall x, y \in \mathbb{R}^n : \quad \sum_{k=1}^{n} |x_k|^\lambda |y_k|^{1-\lambda} \leq \left(\sum_{k=1}^{n} |x_k| \right)^\lambda \left(\sum_{k=1}^{n} |y_k| \right)^{1-\lambda}.$$

These two equivalent inequalities are (both) called Hölder's inequality. The analogue for functions (also called Hölder's inequality) states that for f, g defined on $\mathbb{R}^n$, we have

$$\forall p > 1: \quad \int_{\mathbb{R}^n} |fg| \leq \left(\int_{\mathbb{R}^n} |f|^p \right)^{\frac{1}{p}} \left(\int_{\mathbb{R}^n} |g|^q \right)^{\frac{1}{q}}, \tag{6.20}$$

$$\forall \lambda \in [0,1]: \quad \int_{\mathbb{R}^n} |f|^\lambda |g|^{1-\lambda} \leq \left(\int_{\mathbb{R}^n} |f| \right)^\lambda \left(\int_{\mathbb{R}^n} |g| \right)^{1-\lambda}. \tag{6.21}$$

6.12. Use (6.21) to show that the Euler gamma function Γ defined by (A.6) is log-convex. In fact, a famous result, the Bohr–Mollerup theorem (see Theorem 2.1 in [1]), asserts that this characterizes the Γ function among all functions f satisfying $f(1) = 1$ and $f(x+1) = xf(x)$ for all $x > 0$.

6.13. Here is how the gamma function may be wrapped up in a family. The quantity $\Gamma(\alpha)$ is defined in (A.6) as an integral; you probably noticed that in the last problem, the exponential term in the integrand did not play much of a role; can you prove the following two generalizations?

1. If ψ is a continuous positive function defined on $[0, \infty)$, then the function Φ defined by

$$\Phi(x) := \int_0^\infty \psi(t)t^x \, dt$$

is log-convex.

2. Let ϕ be a continuous positive function defined on $[0, \infty) \times D$ for some convex subset D of $\mathbb{R}^n$. Assume that for all $t > 0$, the map $\phi(t, \cdot)$ is log-convex. Then the function Φ defined by

$$\Phi(x) := \int_0^\infty \phi(t, x) \, dt$$

is log-convex (for more properties of the gamma function, and more generally log-convex functions defined by integrals, see [1]).

Note that the first point can be translated into probabilistic language: if ψ is a probability density, then its rth moment is a log-convex function of r.

6.14. As a prelude to Problem 6.15 let us examine a generalization of the Cauchy–Schwarz inequality. Let ϕ be any nonnegative function; if we take $p = 2$ in (6.20) and replace u, v with $u\sqrt{\phi}, v\sqrt{\phi}$, we obtain the well-known Cauchy–Schwarz inequality:

$$\int |uv|\phi \le \left(\int |u|^2 \phi \right)^{\frac{1}{2}} \left(\int |v|^2 \phi \right)^{\frac{1}{2}}. \tag{6.22}$$

1. If the nonnegative function ϕ has integral 1, show that

$$\left[\int uv\phi - \int u\phi \int v\phi \right]^2 \le \left[\int u^2\phi - \left(\int u\phi \right)^2 \right]\left[\int v^2\phi - \left(\int v\phi \right)^2 \right]. \tag{6.23}$$

2. Deduce that for every function b, the following inequality (in which B denotes the integral of b) holds:

$$\left[B \int uvb - \int ub \int vb \right]^2 \le \left[B \int u^2b - \left(\int ub \right)^2 \right]\left[B \int v^2b - \left(\int vb \right)^2 \right]. \tag{6.24}$$

6.15. It follows from Problem 6.13 that the Euler beta function B is log-convex, but here we are going to prove this by a direct examination of the Hessian.

1. Setting $F := \ln B$, compute the matrix $\nabla^2 F$.
2. Use (6.23) to show that the trace of $\nabla^2 F$ is nonnegative.
3. Use (6.24) to show that the determinant of $\nabla^2 F$ is nonnegative.

6.16. In case you did not enjoy the computations in the previous problem, you might find this more enjoyable; it is well known that the beta and gamma functions are related as follows:

$$B(\alpha, \beta) = \frac{\Gamma(\alpha)\Gamma(\beta)}{\Gamma(\alpha + \beta)}.$$

Use this relation to show that log-convexity of Γ directly implies log-convexity of B.

6.17. This is about sums of log-convex functions.

1. Let g and h be two $\mathcal{C}^2$ convex functions, and define a function f on $D(f) \cap D(g)$ by
$$f := \ln(e^g + e^f).$$

Show that
$$f'' \geq \frac{(g' + h')^2 e^{g+h}}{(e^g + e^h)^2}.$$

2. Deduce that the sum of any finite number of $\mathcal{C}^2$ log-convex functions also is log-convex.

3. Compare your result to what we saw in Problem 6.3 about the LogSumExp function.

The basic quantities of information theory

Summary. In this chapter we introduce the basic quantities that are needed in the implementation of the "entropy dissipation method" for the study of the asymptotics of Markov chains, which will be presented in Section 7.4. This is a very minimal introduction to information theory; for more details we strongly recommend consulting [2, 7, 16, 22].

7.1 Entropy discrete random variables

7.1.1 Entropy of one variable

Definition 7.1. *The entropy of a probability vector $p \in \mathcal{P}_n$ is the following quantity:*

$$H(p) := -\sum_{x \in \mathcal{S}} p(x) \log p(x),$$

with the usual convention $0 \log 0 = 0$.

By extension, if X is a random variable distributed according to p, this will also be called the *entropy of X*, so we will use interchangeably the notation $H(X)$ and $H(p)$. One remark about logarithms and their bases: if the base of the logarithm in the definition above is changed, then the entropy is multiplied by a positive constant; since it is essentially extremal properties that will play a role more than actual numerical values, this is of no consequence. Therefore, everything that will be said henceforth remains true whatever the base of the logarithms; only in applications to binary coding will we have to use logarithms to the base 2, which will be indicated by the notation $\log_2$.

The raison d'être for H is given by the following result:

Theorem 7.2 (Extremal properties of entropy).

Let N be the number of points in $\mathcal{S}$; for a probability vector $p \in \mathcal{P}_N$ we have

$$0 \leq H(X) \leq \log N.$$

© Springer International Publishing AG, part of Springer Nature 2018

J.-F. Collet, *Discrete Stochastic Processes and Applications,*

Universitext, https://doi.org/10.1007/978-3-319-74018-8_7

7   The basic quantities of information theory

*Moreover, the minimum value 0 is attained if and only if p defines a deter-
ministic probability distribution, which means that one component of p is 1
(in which case all other components have to vanish), and the maximum value
$\log N$ is attained if and only if p defines a uniform probability distribution,
which means that all components are $\frac{1}{N}$.*

Proof. Nonnegativity is obvious; for the upper bound it suffices to use the
basic convexity inequality (6.4) with $\phi(z) := z \log z$ and $\lambda_i := \frac{1}{N}$ for all i. If
$H(p) = 0$, all terms in the sum defining H, being of same sign, must vanish, so
for each x, the component $p(x)$ must be either 0 or 1; since p is a probability
vector, we see that exactly one component must be equal to 1. As a matter
of fact, given the strict convexity of ϕ on the interval $(0, 1)$, one sees that the
maximum value may not be attained elsewhere.

This clarifies the role of H: within the range $[0, \log N]$, the numerical value of
$H(p)$ is a measure of "randomness" that tells us which of the two extremes
(deterministic or uniform) we are closer to. In layman's terms, a random
variable is "more unpredictable" if its entropy is close to the maximum value.

   This is reminiscent of another familiar quantity from basic probability, namely
variance. Entropy, however, is more intrinsic to the probability vector
p, in that it depends only on its components and not on what the corresponding points x are: it is well defined if the random variable assumes nonnumerical
values—political parties, vegetables, letters,...—whereas variance can be
defined only for a random variable assuming numerical values and is unit-
dependent. Perhaps the simplest example is coin tossing: if both faces of a coin
carry numerical values $a, b,$ then the variance depends on these values, whereas
entropy does not (which is why strictly speaking, we should talk about the entropy
of p and not of X). Writing $p := P(X = a)$, you may check that

$$var(X) = (a - b)^2 p(1 - p),$$
$$H = p \log p + (1 - p) \log(1 - p),$$

and you may check using elementary calculus that the extremality conditions
for $H(X)$ and $var(X)$ are exactly the same:

$$argmax H(X) = argmax var(X) = \frac{1}{2}. \tag{7.1}$$

Both variance and entropy agree on the fact that the most random case is the
case $p = \frac{1}{2}$, but the maximal value for $var(X)$ depends on a and b. Finally,
note that H is unchanged if we change p into $1 - p$; more generally, if p is a
probability vector, then every probability vector obtained from p by applying
a permutation to its components will have the same entropy.

7.1.2 Entropy of a pair of random variables, conditional entropy

Given two random variables X and Y (on two possibly different spaces $\mathcal{S}$ and
$\mathcal{T}$), one may wonder how the entropy of the pair (X, Y) is related to $H(X)$
and $H(Y)$. Let us compute it by conditioning:

$$H(X,Y) = -\sum_{x \in S}\sum_{y \in T} P(X = x \cap Y = y) \log P(X = x \cap Y = y)$$

$$= -\sum_{x \in S}\sum_{y \in T} P(Y = y|X = x)P(X = x) \log[P(Y = y|X = x)P(X = x)]$$

$$= -\sum_{x \in S} P(X = x) \log P(X = x) \sum_{y \in T} P(Y = y|X = x)$$

$$-\sum_{x \in S} P(X = x) \sum_{y \in T} P(Y = y|X = x) \log P(Y = y|X = x).$$

The penultimate sum is recognized as $H(X)$; the last one resembles the definition of $H(Y)$ in which the distribution of $P(Y = \cdot)$ is replaced by the conditioned version $P(Y = \cdot|X = x)$ and then weighted by $P(X = x)$. We will therefore call it *conditional entropy*:

Definition 7.3. *The entropy of Y conditional on X is the following quantity:*

$$H(Y|X) := -\sum_{x \in S} P(X = x) \sum_{y \in T} P(Y = y|X = x) \log P(Y = y|X = x).$$

Note that in the computation above we could have interchanged the roles of X and Y, so since $H(X,Y) = H(Y,X)$ (we noted earlier that a permutation of states will not change entropy), we obtain

$$H(X,Y) = H(X) + H(Y|X) = H(Y) + H(X|Y). \qquad (7.2)$$

Conditional entropy is subject to a priori inequalities just like ordinary entropy:

Proposition 7.4. *For two random variables X, Y we always have $H(X|Y) \leq H(X)$, with equality when X and Y are independent.*

Proof. The proof just mimics that of the upper bound for $H(X)$: for fixed x put $\lambda_y := P(Y = y)$ and $z_y := P(X = x|Y = y)$. Then inequality (6.4) with the same function ϕ as above gives

$$\sum_{y \in S} P(Y = y)P(Y = y|X = x) \log P(Y = y|X = x) \geq P(X = x) \log P(X = x).$$

The conclusion follows by summation over x. $\quad\square$

Remarks:

- By taking a close look at the inequalities and using strict convexity one can show that the equality $H(X|Y) = H(X)$ in fact characterizes independence; we choose to show this later using another expression for the difference $H(X) - H(X|Y)$ (as a Kullback–Leibler divergence; see below).

- Just as we replaced a variable by a couple of variables when we computed $H(X, Y)$, one may replace Y by a couple of variables in the definition of $H(X|Y)$, and the proof we just gave may be repeated to show that for instance,

$$H(X|Y, Z) \leq H(X|Y).$$

In other words, adding conditioning reduces entropy. Here is the precise general statement, which you should have no difficulty proving:

Proposition 7.5. *If $\mathcal{A}$ and $\mathcal{B}$ are two sets of random variables with $\mathcal{A} \subset \mathcal{B}$, then for a random variable X the following inequality holds:*

$$H(X|\mathcal{B}) \leq \mathcal{H}(\mathcal{X}|\mathcal{A}). \tag{7.3}$$

Intuitively, just as $H(X)$ has the virtues of variance (and better), $H(X|Y)$ has the virtues of covariance, the range now being the interval $[0, H(X)]$. The distance to the upper bound deserves a name:

Definition 7.6. *The mutual information of X and Y is the quantity*

$$I(X, Y) := H(X) - H(X|Y).$$

From what we saw above, it follows that mutual information is always non-negative; returning to (7.2), we see that I is symmetric (hence the adjective *mutual*), which was not the case for conditional entropy, and that if the variables are independent, then their mutual information vanishes.

7.2 Using entropy to count things: Shearer's lemma

The facts that entropy is maximal for the uniform case and that conditioning reduces entropy may be combined to produce interesting estimates of the number of objects in certain finite sets. A good example of this method of "using entropy to count things" (or more precisely to put an upper bound on the number of things) is provided by the famous Loomis–Whitney inequality in geometry. To motivate things, let us begin with the 3-dimensional case (in what follows, the cardinality of a finite set E is denoted by $|E|$):

Theorem 7.7 (The Loomis–Whitney inequality in three dimensions). *Let $\mathcal{X}$, $\mathcal{Y}$, and $\mathcal{Z}$ be three arbitrary sets, and let E be a finite subset of the Cartesian product $\mathcal{X} \times \mathcal{Y} \times \mathcal{Z}$. Let us denote by $E_{\mathcal{X}\mathcal{Y}}$ the projection of E on $\mathcal{X} \times \mathcal{Y}$, and similarly for $E_{\mathcal{X}\mathcal{Z}}$ and $E_{\mathcal{Y}\mathcal{Z}}$. Then*

$$|E|^2 \leq |E_{\mathcal{X}\mathcal{Y}}||E_{\mathcal{X}\mathcal{Z}}||E_{\mathcal{Y}\mathcal{Z}}|. \tag{7.4}$$

Proof. The idea is to use a random variable $U := (X, Y, Z)$ uniformly distributed in E, which therefore has entropy $H(X, Y, Z) = \ln |E|$. The problem is now reduced to finding an upper bound for $H(X, Y, Z)$. The three pairs

(X,Y), (X,Z), and (Y,Z) are the projections of U on respectively E_{XY}, E_{XZ}, and E_{YZ}. These variables may or may not be uniform, but in each case their entropy is bounded from above by the entropy of a uniform variable (for instance $H(X,Y) \leq \ln |E_{XY}|$), whence the idea to try to bound $H(U)$ from above in terms of the entropy of these pairs. Let us begin by splitting $H(U)$ in two different ways:

$$H(X,Y,Z) = H(X|Y,Z) + H(Y,Z) = H(Y|X,Z) + H(X,Z).$$

Adding these two equalities and erasing some of the conditioning to increase entropy, we obtain

$$2H(X,Y,Z) = H(X|Y,Z) + H(Y|X,Z) + H(Y,Z) + H(X,Z)$$
$$\leq H(X|Y) + H(Y) + H(Y,Z) + H(X,Z), \quad (7.5)$$

and therefore

$$2H(X,Y,Z) \leq H(X,Y) + H(Y,Z) + H(X,Z). \quad (7.6)$$

Using the above-mentioned bounds for the three entropies of pairs, this gives

$$2\ln|E| \leq \ln|E_{XY}| + \ln|E_{XZ}| + \ln|E_{YZ}|,$$

which is equivalent to (7.4). □

It is interesting to look at the cases of equality and strict inequality; by drawing a few pictures in $\mathbb{N}^3$ you will see that information is lost (i.e., some points will not be seen) in the projection process if some points in E are aligned in some direction parallel to a coordinate axis; however, these points will be recovered on another projection. In the language of tomography, this means that to count points in a finite subset of $\mathbb{R}^3$ by taking pictures, three projections are required.

How do we go from dimension 3 to dimension n, i.e., how can we bound the entropy of a random vector by sums of entropies of some of its projections? Let us introduce some notation first: if $X := (X_1, \ldots, X_n)$ is a random vector of size n and $A = \{i_1, \ldots, i_k\}$ is a subset of $\{1, \ldots, n\}$, we write $X_A := (X_{i_1}, \ldots, X_{i_k})$. Our generalization is provided by the following beautiful combinatorial result:

Theorem 7.8 (Shearer's lemma). *Let $\mathcal{A}$ be a family of subsets of $\{1, \ldots, n\}$, and let k be an integer such that each point $i \in \{1, \ldots, n\}$ belongs to at least k members of $\mathcal{A}$. Then for every random vector $(X_1, \ldots, X_n)$ we have*

$$H(X_1, \ldots, X_n) \leq \frac{1}{k} \sum_{A \in \mathcal{A}} H(X_A).$$

If we take for $\mathcal{A}$ the set of singletons, then $k = 1$, and the conclusion of Shearer's lemma is just subadditivity:

$$H(X_1, \ldots, X_n) \leq H(X_1) + \cdots + H(X_n).$$

If now we take for $\mathcal{A}$ the set of pairs, we have $k = n - 1$; if $n = 3$, we exactly recover inequality (7.6).

Proof. Pick an element $A = \{i_1, \ldots, i_r\}$ of $\mathcal{A}$; using the chain rule for entropy, we have

$$H(X_A) = H(X_{i_r}|X_{i_{r-1}}, \ldots, X_{i_1}) + \cdots + H(X_{i_2}|X_{i_1}) + H(X_{i_1}).$$

If we add some conditioning, this decreases entropy:

$$H(X_A) \geq H(X_{i_r}|X_{i_{r-1}}, \ldots, X_1) + \cdots + H(X_{i_2}|X_{i_1}, \ldots X_1) + H(X_{i_1}).$$

In this last sum the notation $H(X_j|X_l, \ldots, X_1)$ means that the conditioner now includes all indices between 1 and l, and not just those present in A. In other words, we have

$$H(X_A) \geq \sum_{j \in A} H(X_j|X_{j-1}, \ldots, X_1).$$

If we now sum over $A \in \mathcal{A}$, each term appears at least k times; since all terms are nonnegative, we obtain the lower bound

$$\sum_{A \in \mathcal{A}} H(X_A) \geq k \sum_{j=1}^{n} H(X_j|X_{j-1}, \ldots, X_1),$$

which (by the chain rule for entropy again) is just $kH(X)$. $\square$

7.3 Φ-divergences

As announced in the introduction, information theory makes extensive use of some quantities that while not being distances in the true mathematical sense, do retain some properties of true distances. These quantities are all constructed from convex functions, and their properties will always be consequences of the convexity inequalities we saw in the previous chapter.

7.3.1 General form

Let us begin with a general definition:

Definition 7.9. *Let Φ be a convex map defined on $[0, \infty)$. The Φ-divergence is the map D_Φ defined on $\mathcal{P}_n \times \mathcal{P}_n^*$ as follows:*

$$D_\Phi(p, q) := \sum_{x \in \mathcal{S}} q(x)\Phi\left(\frac{p(x)}{q(x)}\right).$$

The following property, sometimes referred to as Gibbs's lemma (sometimes the name Gibbs's lemma is used only for the case $\phi(z) = z \log z$), shows the utility of D_Φ:

Lemma 7.10 (Gibbs's lemma). *Assume that Φ is strictly convex; then*

- $\forall p, q \in \mathcal{P}_n \times \mathcal{P}_n^*$: $D_\Phi(p, q) \geq \Phi(1)$, *and*
- $D_\Phi(p, q) = \Phi(1) \Leftrightarrow p = q$.

This gives us a criterion for identifying p and q (most of the time we will take $\Phi(1) = 0$). In the introduction to the chapter on convex analysis we mentioned that multiplication of a probability vector by a Markov matrix has a regularizing effect; we now have the machinery to give a precise statement:

Theorem 7.11. *Let P be an $n \times n$ stochastic matrix and $p, q \in \mathcal{P}_n \times \mathcal{P}_n^*$; then for every differentiable function ϕ, one has*

$$D_\phi(p, q) - D_\phi(pP, qP) = \sum_x \sum_y q(x) P(x, y) B_\phi \left(\frac{p(x)}{q(x)}, \frac{(pP)(y)}{(qP)(y)} \right). \qquad (7.7)$$

Proof. As we did before in the proof of (6.15), we begin by writing the difference of the sums as a unique double sum, and then rewrite the general term as a Bregman difference:

$$D_\phi(p, q) - D_\phi(pP, qP) = \sum_x q(x) \phi \left(\frac{p(x)}{q(x)} \right) - \sum_y (qP)(y) \phi \left(\frac{(pP)(y)}{(qP)(y)} \right)$$

$$= \sum_x \sum_y q(x) P(x, y) \left[\phi \left(\frac{p(x)}{q(x)} \right) - \phi \left(\frac{(pP)(y)}{(qP)(y)} \right) \right]$$

$$= \sum_x \sum_y q(x) P(x, y) B_\phi \left(\frac{p(x)}{q(x)}, \frac{(pP)(y)}{(qP)(y)} \right)$$

$$+ \sum_x \sum_y q(x) P(x, y) \phi' \left(\frac{(pP)(y)}{(qP)(y)} \right) \left[\frac{p(x)}{q(x)} - \frac{(pP)(y)}{(qP)(y)} \right].$$

Owing to the definition of $(pP)(y)$ and $(qP)(y)$, the last sum involving ϕ' vanishes (exactly as it did in the proof of (6.45)), which establishes (7.7).

7.3.2 The most famous Φ-Divergence: Kullback–Leibler divergence

The particular choice $\Phi(z) = z \log z$ is probably the most widely used (besides being historically the first), and it yields what is now called the *Kullback–Leibler divergence*:

Definition 7.12. *The Kullback–Leibler divergence between* $p \in \mathcal{P}_n$ *and* $q \in \mathcal{P}_n^*$ *is defined as follows:*

$$D_{KL}(p,q) := \sum_{x \in S} p(x) \log(\frac{p(x)}{q(x)})$$

(as above in the definition of entropy, the logarithms may be taken to any base greater than 1). Beware! This is not truly speaking a distance: it is not symmetric, and the triangle inequality fails; it is, however, quite useful, mainly for the following reason:

$$\forall p, q \in \mathcal{P}_n \times \mathcal{P}_n^* : D_{KL}(p,q) \geq 0; \ D_{KL}(p,q) = 0 \Leftrightarrow p = q$$

(note an amusing analogy with travel fares: the return is not necessarily priced the same as the original trip; there is no triangle inequality; but the only way to spend nothing is to stay home).

Now if we take q uniformly distributed, we obtain

$$D_{KL}(p,q) = \log n - H(p) = H(q) - H(p),$$

from which we recover the by now familiar fact that $H(p)$ is maximal if and only if p is uniform. This turns out to be one special instance of a more general situation that may be roughly described as follows: in a given class $\mathcal{M} \subset \mathcal{P}_n$ of probability vectors (for instance, probability vectors with a certain number of moments prescribed), there is a unique element q satisfying

$$\forall p \in \mathcal{M} : \quad D_{KL}(p,q) = H(q) - H(p),$$

from which it follows that q is the unique maximizer of H on $\mathcal{M}$. This will be further developed in the section on exponential families (see in particular Theorem 7.22).

Although D_{KL} is not a distance, it may serve the purpose of a distance in the study of convergence; more precisely, if $D_{KL}(p,q)$ is small, then p and q are close in the l^1 norm. Let us just recall that the l^1 norm of a vector $p \in \mathbb{R}^n$ is defined as

$$|p|_1 := \sum_x |p(x)|.$$

It is a classical fact that $|\cdot|_1$ is indeed a norm on $\mathbb{R}^n$; we shall here use the restriction to $\mathcal{P}_n \times \mathcal{P}_n^*$ of the induced distance. Here is the precise statement of how D_{KL} "controls" the l^1 distance from above:

Theorem 7.13 (Pinsker's inequality). *For all* $p, q \in \mathcal{P}_n \times \mathcal{P}_n^*$ *we have*

$$D_{KL}(p,q) \geq \frac{1}{2}|p - q|_1^2. \tag{7.8}$$

The proof relies on the so-called *log sum inequality:*

Lemma 7.14 (log sum inequality). *Let $p, q \in \mathbb{R}^n$ be two vectors with positive entries, $p(x), q(x) > 0$, and set*

$$P := \sum_x p(x), \ Q := \sum_x q(x).$$

Then we have

$$\sum_x p(x) \log \frac{p(x)}{q(x)} \geq P \log \frac{P}{Q}.$$

Let us emphasize that here p and q are not supposed to be probability vectors; the proof mimics the proof of Gibbs's lemma given above (in fact, the dedicated reader, having done Problem 6.3, has already proved this inequality by another method): setting $\phi(z) = z \log z$ and using the convexity of ϕ, we have

$$\sum_x p(x) \log \frac{p(x)}{q(x)} = Q \sum_x \frac{q(x)}{Q} \phi\left(\frac{p(x)}{q(x)}\right)$$

$$\geq Q\phi\left(\sum_x \frac{p(x)}{Q}\right) = P \log \frac{P}{Q}.$$

With this in hand we may now give the proof of Pinsker's inequality. Let us begin with the case $n = 2$ and show that the inequality is true for the case of two Bernoulli variables. Calling p and q their parameters, the goal is to show that for $0 < p, q < 1$, we have

$$p \log \frac{p}{q} + (1 - p) \log \frac{1 - p}{1 - q} - 2(p - q)^2 \geq 0. \tag{7.9}$$

Let q be fixed; considering the quantity above as a function $g(p)$, we may easily compute the two derivatives:

$$g'(p) = \log\left(\frac{p}{1 - p} \frac{1 - q}{q}\right) - 4(p - q), \ g''(p) = \frac{1 - 4p(1 - p)}{p(1 - p)}.$$

The term $p(1 - p)$ is obviously less than $\frac{1}{4}$ (interestingly enough, this is the variance of our Bernoulli variable, which we met in (7.1)). Therefore, g is convex, but since $g'(q)$ vanishes, we see that g is minimal at q, where it is zero, which establishes (7.9). For the general case let us split the sums defining $D_{KL}(p, q)$ and $|p - q|_1$ according to whether $p(x)$ is greater than or less than $q(x)$. Setting

$$P_1 := \sum_{p(x) \geq q(x)} p(x), \ P_2 := \sum_{p(x) < q(x)} p(x) = 1 - P_1,$$

$$Q_1 := \sum_{p(x) \geq q(x)} q(x), \ Q_2 := \sum_{p(x) < q(x)} q(x) = 1 - Q_1,$$

we obtain

$$|p - q|_1 = P_1 - P_2 - Q_1 + Q_2 = 2(P_1 - Q_1), \qquad (7.10)$$

$$D_{KL}(p, q) = \sum_{p(x) \geq q(x)} p(x) \log \frac{p(x)}{q(x)} + \sum_{p(x) < q(x)} p(x) \log \frac{p(x)}{q(x)}. \qquad (7.11)$$

Applying the log sum inequality to (7.11), we obtain

$$D_{KL}(p, q) \geq P_1 \log \frac{P_1}{Q_1} + P_2 \log \frac{P_2}{Q_2}.$$

We are therefore back to the case $n = 2$, and we may apply (7.9) to get

$$D_{KL}(p, q) \geq 2(P_1 - Q_1)^2,$$

which in view of (7.10), gives (7.8).   □

In a way, Pinsker's inequality may be thought of as a quantitative refinement of Gibbs's lemma; in particular, it shows that sequences (p_n) of probability vectors that "converge" in the sense of D_{KL} indeed converge in the l^1 metric:

$$D_{KL}(p_n, q_n) \to 0 \Rightarrow |p_n - q_n|_1 \to 0.$$

This will be used in particular in the study of the asymptotic behavior of Markov processes in the next section.

Let us now move on to an interesting use of D_{KL} to provide some sort of "distance to independence." Let X and Y be two variables with probability vectors p_X and p_Y, and call $p_{X,Y}$ the probability vector of the pair (X, Y). As we all know, if these variables are independent, we have $p_{X,Y}(x, y) = p_X(x)p_Y(y)$ for all (x, y). In the general case, now let us define the vector $p_X \otimes p_Y$ by

$$\forall x, y : \quad (p_X \otimes p_Y)(x, y) = p_X(x)p_Y(y).$$

In other words, $p_X \otimes p_Y$ is *what the vector* $p_{X,Y}$ *would be* were the variables X and Y independent. The idea is now that the quantity $D_{KL}(p_{X,Y}, p_X \otimes p_Y)$ should provide a measure of the level of dependency of X and Y.

It is very easy to check by direct computation that the following relation (which is, in fact, in many textbooks taken as the definition for $I(X, Y)$) holds:

Lemma 7.15.

$$D_{KL}(p_{X,Y}, p_X \otimes p_Y) = I(X, Y).$$

This relation immediately gives the characterization of independence that we announced earlier:

Proposition 7.16. *Two variables* X *and* Y *are independent if and only if* $I(X, Y) = 0$.

7.4 Entropy dissipation in Markov processes

In Chapter 1 we proved that an ergodic chain has a unique stationary distribution π, and that the probability distribution of the chain converges to π for large times. The method of proof, relying on the coupling method, was essentially probabilistic; then in Chapter 2, a linear-algebraic approach to the same result was implemented; it turns out that the property of entropy dissipation under multiplication by a Markov matrix (7.7) makes it possible to give a quick and purely analytic proof of the convergence statement in Theorem 1.39. To avoid technicalities involved with infinite sums, we restrict ourselves to the case of a finite space:

Theorem 7.17. *Let P be an ergodic Markov matrix on a finite space $\mathcal{S}$:*

$$\forall n \geq n_0, \forall x, y \in \mathcal{S} : P^n(x, y) > 0.$$

As we know from the first half of the conclusion in Theorem 1.39, there exists a unique stationary distribution π, and all its components are nonzero. Then for every probability vector p we have

$$\lim_{n \to \infty} (pP^n)(x) = \pi(x).$$

Proof. Let ϕ be any differentiable convex function. Using (7.7) with pP^n in place of p, and π in place of q, we have

$$D_\phi(pP^n, \pi) - D_\phi(pP^{n+1}, \pi)$$

$$= \sum_x \sum_y \pi(x) P^{n+1}(x, y) \phi\left(\frac{(pP^n)(x)}{\pi(x)}\right) - \phi\left(\frac{(pP^{n+1})(y)}{\pi(y)}\right)$$

$$= \sum_x \sum_y \pi(x) P^{n+1}(x, y) B_\phi\left(\frac{(pP^n)(x)}{\pi(x)}, \frac{(pP^{n+1})(y)}{\pi(y)}\right) \geq 0.$$

Setting $u_n := D_\phi(pP^n, \pi)$, this shows that the sequence (u_n) is nonincreasing, but since it is bounded from below by $\phi(1)$, it has to converge to some limit; therefore, the difference $u_n - u_{n+1}$ must converge to 0, and so (being all nonnegative) all terms in the last sum above must also converge to 0:

$$\forall x, y : B_\phi\left(\frac{(pP^n)(x)}{\pi(x)}, \frac{(pP^{n+1})(y)}{\pi(y)}\right) \to 0.$$

This is true for every function ϕ; choosing $\phi(z) = z^2$ (or any other function whose second derivative is bounded away from zero), we get

$$\frac{(pP^n)(x)}{\pi(x)} - \frac{(pP^{n+1})(y)}{\pi(y)} \to 0.$$

Multiplying by $\pi(y)$ and then summing over y, we get the convergence result. $\square$

Remark: if the matrix is not ergodic (so that there may not even exist any stationary distribution), from (7.7) we see that the associated Markov chain still has the following interesting dissipation property:

$$D_\phi(pP^n, qP^n) - D_\phi(pP^{n+1}, qP^{n+1})$$

$$= \sum_x \sum_y q(x)P^{n+1}(x,y)B_\phi\left(\frac{(pP^n)(x)}{(qP^n)(x)}, \frac{(pP^{n+1})(y)}{(qP^{n+1})(y)}\right) \geq 0,$$

valid for any two probability vectors p, q. In other words, the two orbits originating from p and q tend to come close in ϕ-divergence.

Markov chains in continuous time also are dissipative; more precisely, the time derivative of the ϕ-divergence between two orbits may be expressed as a Bregman divergence (to avoid technicalities, we give here only the statement for finite spaces; also for a related identity, see Problem 7.10):

Theorem 7.18. *Let Q be an $n \times n$ rate matrix, and assume that the associated Markov chain is irreducible. For $t > 0$ let $p_t, q_t \in \mathcal{P}_n$ be two solutions to the associated backward Kolmogorov equation:*

$$\forall t > 0 : \quad \frac{d}{dt}p_t = Qp_t, \; \frac{d}{dt}q_t = Qq_t.$$

Then for every differentiable function ϕ defined on $[0, \infty)$, we have

$$\frac{d}{dt}D_\phi(p_t, q_t) = -\sum_x \sum_y q_t(y)Q(y,x)B_\phi\left(\frac{p_t(y)}{q_t(y)}, \frac{p_t(x)}{q_t(x)}\right). \tag{7.12}$$

If the chain admits a reversible distribution π, we have

$$\frac{d}{dt}D_\phi(p_t, \pi) = -\frac{1}{2}\sum_x \sum_y Q(y,x)\pi(y)\left[\frac{p_t(y)}{q_t(y)} - \frac{p_t(x)}{q_t(x)}\right]\left[\phi'\left(\frac{p_t(y)}{q_t(y)}\right) - \phi'\left(\frac{p_t(x)}{q_t(x)}\right)\right].$$

$$\tag{7.13}$$

Proof. First note that since the chain is irreducible, by Theorem 4.26 we have $q_t(x) > 0$ for all x and all $t > 0$, which implies that $D_\phi(p_t, q_t)$ is well defined for all $t > 0$; also we have $\pi(x) > 0$ for all x. By straightforward differentiation we have

$$\frac{d}{dt}D_\phi(p_t, q_t) = \sum_x \left[q_t'(x)\phi\left(\frac{p_t(x)}{q_t(x)}\right) + \left(p_t'(x) - \frac{p_t(x)}{q_t(x)}q_t'(x)\right)\phi'\left(\frac{p_t(x)}{q_t(x)}\right)\right].$$

Thus taking into account the Kolmogorov equations, we obtain

$$\frac{d}{dt}D_\phi(p_t, q_t) = \sum_x \sum_y Q(y,x)q_t(y)\left[\phi\left(\frac{p_t(x)}{q_t(x)}\right) + \left(\frac{p_t(y)}{q_t(y)} - \frac{p_t(x)}{q_t(x)}\right)\phi'\left(\frac{p_t(x)}{q_t(x)}\right)\right].$$

$$\tag{7.14}$$

Rewriting the term in brackets as a Bregman divergence, we get

$$\frac{d}{dt} D_\phi(p_t, q_t) = \sum_x \sum_y Q(y,x) q_t(y) [\phi(\frac{p_t(y)}{q_t(y)}) - B_\phi(\frac{p_t(y)}{q_t(y)}; \frac{p_t(x)}{q_t(x)})],$$

which yields (7.12), since Q is a rate matrix. Now if we have a stationary distribution π, taking $q = \pi$ in (7.14), we have

$$\frac{d}{dt} D_\phi(p_t, \pi) = \sum_x \sum_y Q(y,x) \pi(y) (\frac{p_t(y)}{\pi(y)} - \frac{p_t(x)}{\pi(x)}) \phi'(\frac{p_t(x)}{\pi(x)}).$$

Call this last sum S_1; if we symmetrize S_1, i.e., if we swap x and y in its definition, we obtain a sum S_2, which is in fact equal to S_1. Using the local balance relation $Q(y,x)\pi(y) = Q(x,y)\pi(x)$ and the equality $S_1 = \frac{S_1+S_2}{2}$, we obtain (7.13). $\square$

7.5 Exponential families

Exponential families are a class of parametric distributions for which the quantities introduced above lend themselves to particularly simple computations; this is probably what makes them so useful in classification problems. One way to motivate the definition of exponential families is to look for distributions that maximize entropy under some constraints (although historically, this is not how they were introduced, some fifty years earlier than the formalization of the entropy maximization principle; for an introduction to exponential families related to the theory of point estimation, see, for instance, [21] and [20]). Let us begin with some (not totally rigorous) heuristics: if t is a given function from $[0, \infty)$ into $\mathbb{R}^r$, let us look for a probability distribution over some given space that maximizes $H(p)$ subject to the constraint that $E_p(t)$ is fixed. For instance, if $t(x) = x$, then we are looking for an entropy-maximizing distribution with prescribed expectation. For the sake of clarity we will work with absolutely continuous distributions, but our reasoning would apply to discrete variables as well if we replaced the integrals by sums. If $t = (t_1, \ldots, t_r)$ and $m := (m_1, \ldots, m_r)$ is a collection of r given numbers, call $\mathcal{M}_m$ the set of probability distributions f satisfying the constraints

$$\int t_1(x) f(x)\, dx = m_1, \ldots, \int t_r(x) f(x)\, dx = m_r.$$

The goal is to find $f \in \mathcal{M}_m$ such that

$$\forall g \in \mathcal{M}_m : \quad H(f) \geq H(g).$$

This sounds like a reasonable quest, since H is concave. Now, since the constraints are linear, we are dealing with a standard constrained optimization problem for which we may use Lagrange multipliers (if you are not familiar with Lagrange multipliers, you may skip this motivational discussion—after all, it is only formal—and jump to the formal definition). To simplify the computation, note that maximizing $H(f)$ on $\mathcal{M}_m$ is equivalent to maximizing $H(f) + \int f$ on the same set. The associated Lagrangian function L is defined on $\mathcal{M}_m \times \mathbb{R}^{r+1}$ by

$$L(f, \lambda_0, \cdot \lambda_r) := - \int f(x) \ln f(x)\, dx$$

$$+\lambda_0 \left(\int f(x)\, dx - 1 \right) + \lambda_1 \left(\int t_1(x) f(x)\, dx - m_1 \right) + \cdots + \lambda_r \left(\int f(x)\, dx - m_r \right).$$

$$(7.15)$$

Now let h be a function such that

$$\int h(x)\, dx = 0, \int t_1(x) h(x)\, dx = 0, \ldots, \int t_r(x) h(x)\, dx = 0.$$

Since all the above integrals vanish, h (as well as any multiple of h thanks to the linearity of the constraints) can be used as an increment in the argument of L, that is, for all $t > 0$, the function $f + th$ is a member of $\mathcal{M}_m$, and the quantity $L(f + th)$ is maximal at $t = 0$. Therefore, the so-called *first variation* must vanish:

$$\frac{d}{dt} L(f + th) = \int [\ln f - \lambda_0 - \lambda_1 t_1(x) - \cdots - \lambda_r t_r(x)] h(x)\, dx = 0.$$

Since this is true for "many" functions h, the bracket in the integrand must vanish, and we get an expression for the optimal distribution:

$$f(x) = \exp(\lambda_0 + \lambda_1 t_1(x) + \cdots + \lambda_r t_r(x)). \quad \square$$

Note that we do not really know (or care) what the optimal λ_i are, but let us just retain the qualitative conclusion that on formal grounds, the optimal f is the exponential of a linear combination of the constraints. It is rather easy to dot all the mathematical i's in this informal discussion, but in fact, we will work backward and get the optimality result as a proper theorem about exponential families, once we have defined them properly. Adopting the notational standard of statistics (f becomes p, and λ, standard code name for Lagrange multipliers, becomes θ, standard code name for parameters in statistics), we would like to define an exponential family as the parametric family

$$p_\theta(x) = \exp(t(x) \cdot \theta - F(\theta) + k(x)).$$

At this point it is not really clear why the term $k(x)$ is added; it will enable us to include more classical distributions than just the entropy maximizers in our class. Let us emphasize that p_θ is a scalar distribution, so that x varies in some subset of $\mathbb{R}$, whereas $t(x) \in \mathbb{R}^p$. Note that by a slight abuse of notation (and even though we did our motivational discussion using integrals), we include here discrete as well as continuous distributions: if we are dealing with a discrete variable X_θ, here $p_\theta(x)$ stands for the quantity $P(X_\theta = x)$, whereas if X_θ is a continuous variable, $p_\theta(x)$ stands for its density. The notation used in the definition seems to be the most common, although it has not been quite universally adopted. Prescribing this form for p_θ raises two issues: integrability (or summability if the distribution is discrete), and integrability to 1. Formally, if p_θ is a probability distribution, we must have

$$e^{-F(\theta)} \int \exp\left(t(x)\theta + k(x)\right) dx = 1 \qquad (7.16)$$

for a continuous distribution, or the analogous relation with a sum in place of the integral in the discrete case. This would show that in fact, F is defined in terms of t and k by

$$F(\theta) = \log \int \exp\left(t(x)\theta + k(x)\right) dx, \qquad (7.17)$$

so we would like to define an exponential family by specifying t and k. But how can we make sure that we are manipulating integrable quantities? Switching back to rigorous mathematical mode, the following result tells us exactly when this is the case:

Theorem 7.19 (Existence of exponential families). *Let*

$$t : \mathbb{R} \to \mathbb{R}^p, \quad k : \mathbb{R} \to \mathbb{R}$$

be two functions with the same domain. Then the following hold:

- *The set Θ of all values θ for which the quantity $\exp\left(t(x)\theta + k(x)\right)$ is integrable with respect to x is a convex subset of $\mathbb{R}^p$.*
- *For every integrable function f, the function ψ defined on the interior of Θ by*

$$\psi(\theta) := \int f(x) \exp\left(t(x)\theta + k(x)\right) dx$$

is C^∞, and its derivatives may be computed in the interior of Θ by differentiation under the integral sign.
- *The map F defined by (7.17) is defined on all of Θ. It is convex and C^∞ on the interior of Θ.*

For the proof, see, for instance, Lemma 2.7.1 and Theorem 2.7.1 in [20] (note that the convexity of F is a consequence of Problem 6.13).

We can finally give the proper definition:

Definition 7.20. *Let t and k be as in the theorem, and define the set Θ as in the theorem and the map F on Θ by (7.17). Then for all $\theta \in \Theta$, the function p_θ defined by*

$$p_\theta(x) = \exp\left(t(x) \cdot \theta - F(\theta) + k(x)\right) \qquad (7.18)$$

is a probability distribution. The collection $(p_\theta)_{\theta \in \Theta}$ is called an exponential family, and the set Θ is called its natural parameter space.

7 The basic quantities of information theory

Henceforth X_θ will denote a random variable with probability density (or mass function) p_θ. In fact, F turns out to have a probabilistic significance:

Theorem 7.21. *For all $\theta \in \Theta$, the gradient $\nabla F(\theta)$ is the expectation of the random vector $t(X_\theta)$, and the Hessian $\nabla^2 F(\theta)$ is its covariance matrix:*

$$\frac{\partial F}{\partial \theta_i} = E(t_i(X_\theta)), \quad \frac{\partial^2 F}{\partial \theta_i \partial \theta_j} = cov(t_i(X_\theta), t_j(X_\theta)). \tag{7.19}$$

Moreover, the moment-generating function of $t(X_\theta)$ is given by

$$M_{t(X_\theta)}(u) := E(e^{u \cdot t(X_\theta)}) = \frac{\exp(F(u + \theta))}{\exp(F(\theta))}. \tag{7.20}$$

Proof. The moment-generating function $M_{t(X_\theta)}(u)$ is given by

$$E(e^{u \cdot t(X_\theta)}) = \int \exp(t(x)(u + \theta) - F(\theta) + k(x)) \, dx.$$

Equality (7.20) is readily obtained by combining this relation with (7.17). It is easy to obtain (7.19) from (7.20) by differentiation (see Problem 7.11), but these two relations may also be obtained by a direct computation: let us first differentiate (7.16) with respect to θ_i. We obtain

$$-\frac{\partial F}{\partial \theta_i} e^{-F(\theta)} \int \exp(t(x)\theta + k(x)) \, dx + e^{-F(\theta)} \int t_i(x) \exp(t(x)\theta + k(x)) \, dx = 0,$$

so that

$$\frac{\partial F}{\partial \theta_i} = e^{-F(\theta)} \int t_i(x) \exp(t(x)\theta + k(x)) \, dx = E(t_i(X_\theta)).$$

We may now differentiate this last relation with respect to θ_j:

$$\frac{\partial^2 F}{\partial \theta_i \partial \theta_j} = -e^{-F(\theta)} \frac{\partial F}{\partial \theta_j} \int t_i(x) \exp(t(x)\theta + k(x)) \, dx$$

$$+ e^{-F(\theta)} \int t_i(x) t_j(x) \exp(t(x)\theta + k(x)) \, dx$$

$$= -E(t_j(X_\theta)) E(t_i(X_\theta)) + E(t_j(X_\theta) t_i(X_\theta)).$$

In fact, this last relation gives yet another proof that (7.17) gives a convex function: the matrix $\nabla^2 F(\theta)$ being the covariance matrix of a random vector, it has to be positive semidefinite.

At this point perhaps a worked example would be in order before we move to the general properties of exponential families. Take the well-known binomial distribution $\mathcal{B}(n, p)$, considering n fixed. By this we mean that the parameter θ will be defined in terms of p only, and n will not be allowed to vary (working some of the distributions in Problem 7.12, you will see that sometimes for a given distribution to fit in the general framework of the exponential family, some parameters have to be kept constant). The strategy for rewriting a given parametric distribution as an exponential family is always the same: take the logarithm of the density (or mass function), and reorder terms so as to have first some linear combination of some functions of the variable x, then something that depends only on the original parameters, and finally something that depends only on x. This then dictates (by order of appearance) the definition of the natural parameter θ in terms of the original one(s) and the expression of $t(x)$ and $F(\theta)$. For the binomial distribution we obtain the form

$$\log P(X = x) = x \log \frac{p}{1 - p} + n \log (1 - p) + \log \binom{n}{x}, \quad 0 \le x \le n,$$

which immediately implies

$$t(x) = x, \ \theta = \log \frac{p}{1 - p}, F(\theta) = -n \log (1 - p) = n \log (1 + e^\theta), k(x) = \log \binom{n}{x}.$$

Note that the expression for F is obtained by inverting the expression for θ in terms of p, which fortunately is explicitly doable for a number of classical distributions. Now that we have F, we can use (7.19) to recover the well-known formula for the expectation and variance of the binomial distribution:

$$E(X) = np, var(X) = np(1 - p)$$

(please do the algebra; you have to compute the derivatives of F and go back to the original parameter p). Note that the Bernoulli distribution is included in what we just did, by taking $n = 1$. In appendix E you will find tables that group most of the classical exponential families (discrete as well as continuous), the natural parameters being indicated in each case. At this point you are all set do Problem 7.12.

In our informal discussion at the beginning of this section we were seeking entropy-maximizing distributions, and this led us to our definition; here, finally, comes the precise maximization theorem:

Theorem 7.22. *Consider the exponential family defined by (7.18). Then the following hold:*

- *For every variable X such that the quantities $H(X)$, $E(t(X))$, and $E(k(X))$ are finite, we have*

$$D_{KL}(X, X_\theta) =$$
$$H(X_\theta) - H(X) + \theta E(t(X_\theta) - t(X)) + E(k(X_\theta) - k(X)). \quad (7.21)$$

- *As a consequence, if k is identically zero, then X_θ is the unique maximizer of H on the set of probability distributions X satisfying $E(t(X) = E(t(X_\theta)$ and $H(X) < \infty$.*

Proof. From (7.18), the entropy of X_θ is

$$H(X_\theta) = -\theta E(t(X_\theta)) + F(\theta) - E(k(X_\theta)). \tag{7.22}$$

On the other hand, we have

$$D_{KL}(X, X_\theta) = \int p_X(x) \log \frac{p_X(x)}{p_\theta(x)}\, dx$$

$$= -H(X) - \int p_X(x)[t(x)\theta - F(\theta) + k(x)]\, dx$$

$$= -H(X) - \theta E(t(X)) + F(\theta) - E(k(X)).$$

Subtracting this last relation from (7.22), we obtain (7.21).   □

A closely related question is how the maximum value for H behaves as the value of the constraint $E(t(X))$ (therefore of the parameter θ) is changed; here again the algebra of exponential families makes the computations particularly simple. More precisely, entropy and Kullback–Leibler divergence have a very simple expression in terms of Legendre transforms and Bregman divergence:

Theorem 7.23. *For all $\theta, \mu \in \Theta$ we have*
$$H(X_\theta) = -F^*(\theta*) - E(k(X_\theta)), \tag{7.23}$$
$$D_{KL}(X_\theta, X_\mu) = B_F(\mu, \theta). \tag{7.24}$$

Proof. The first relation is immediately obtained from (7.22), (7.19), and (6.8); similarly, the second relation is a direct consequence of (7.19):

$$D_{KL}(X_\theta, X_\mu) = \int p_\theta(x) \log \frac{p_\theta(x)}{p_\mu(x)}\, dx$$

$$= \int \exp\left(t(x)\theta - F(\theta) + k(x)\right)[t(x)(\theta - \mu) + F(\mu) - F(\theta)]\, dx$$

$$= F(\mu) - F(\theta) + (\theta - \mu)\int t(x)\exp\left(t(x)\theta - F(\theta) + k(x)\right) dx$$

$$= F(\mu) - F(\theta) + (\theta - \mu)F'(\theta) = B_F(\mu, \theta).$$

□

For many classical distributions, the function k is identically zero (see Problem 7.13); in this case, the formula given above for $H(p_\theta)$ has an interesting consequence: finding θ that maximizes $H(p_\theta)$ is then equivalent to finding θ^* that minimizes $F^*(\theta*)$.

7.6 Problems

7.1. Compute the entropy $H(p)$ of a Bernoulli variable of parameter p and study the variations of this quantity for $p \in [0, 1]$. Can you interpret the result, and can you see a parallel with the variance?

7.2. Here is another way to prove Gibbs's lemma.

1. Show that for every positive x we have $-\log x + x - 1 \geq 0$, with equality if and only if $x = 1$.
2. Using the relation

$$D_{KL}(p, q) = \sum_x p(x)\left(-\log \frac{q(x)}{p(x)} + \frac{q(x)}{p(x)} - 1\right)$$

(why is this true?) deduce that $D_{KL}(p, q) = 0$ if and only if $p = q$.

7.3. A Markov matrix P is said to be *doubly stochastic* if its columns also add up to 1:

$$\forall y : \sum_x P(x, y) = 1.$$

A *permutation matrix* is a doubly stochastic matrix that contains only 0's and 1's (for some vector $p \in \mathbb{R}^n$ please compute the product pP; that will tell you why this is called a *permutation matrix*). A classical result (known as Birkhoff's theorem) asserts that every doubly stochastic matrix is a convex combination of permutation matrices.

1. Show that a Markov matrix is doubly stochastic if and only if it has the uniform distribution as stationary distribution.
2. Show that if P is doubly stochastic, then for every probability vector p, one has $H(pP) \geq H(p)$.
3. Conversely, show that if an $n \times n$ Markov matrix P is such that $H(pP) \geq H(p)$ for all $p \in \mathcal{P}_n$, then P is doubly stochastic.
4. If P is doubly stochastic, show that for every vector p we have

$$\prod_x p(x) \leq \prod_x (pP)(x).$$

7.4. Let us return to inequality (7.4).

1. Can you think of sets for which this in fact an equality? Or sets for which it is strict?
2. Using inequality (6.22), can you state and prove a continuous analogue of this inequality (see [30] for more details and interesting applications to some counting problems in graph theory)?

7.5. In this problem we examine the behavior of entropy under composition by a function.

1. Let X be any discrete random variable, and let f be a map whose domain includes all values of X. For all x and y, see what values the quantity $P(f(X) = y | X = x)$ may assume, and deduce the value of $H(f(X)|X)$.
2. By expressing $H(X, f(X))$ in two different ways, show that $H(f(X)) \leq H(X)$. How can you interpret this inequality?

7 The basic quantities of information theory

3. Let X assume the values $-1, 0, 1$ with respective probabilities $\frac{1}{4}, \frac{1}{2}, \frac{1}{4}$. Compute $H(X)$ and $H(X^2)$.
4. Same question as in the previous problem if X assumes the values $0, 1, 2$ with respective probabilities $\frac{1}{4}, \frac{1}{2}, \frac{1}{4}$. Comments?
5. Now we have understood what is going on, and we can guess that the following statement must hold: if a map f is injective on the image of X, then $H(f(X)) = H(X)$. Can you write down a proof of this? In the next two questions we look at the converse.
6. Let us begin with the simple case that two points $x_1, x_2 \in \mathcal{X}$ satisfy $f(x_1) = f(x_2)$, and f is injective on the set $\mathcal{X} \setminus \{x_1, x_2\}$. Assume that both $p := P(X = x_1)$ and $q = P(X = x_2)$ are nonzero. Express the difference $H(X) - H(f(X))$ in terms of p and q, and deduce that $H(f(X)) < H(X)$.
7. Let now f be any map defined on $\mathcal{X}$. Prove the relation

$$H(X) - H(f(X)) = \sum_{x \in \mathcal{X}} P(X = x) \log \frac{P(f(X) = f(x))}{P(X = x)},$$

and conclude that $H(X) \geq H(f(X))$ (beware, the right hand side is not a Kullback–Leibler divergence).

7.6. Compute the Kullback–Leibler divergence between X and Y in the following cases:

1. Geometric distributions with parameters p and q.
2. Exponential distributions with parameters λ and μ.
3. Gamma distributions with parameters (n, λ) and (n, μ).
4. Binomial distributions with parameters (n, p) and (n, q).
5. Poisson distributions with parameters λ and μ.

In each of the preceding cases set $t := \frac{\mu}{\lambda}$ or $t := \frac{p}{q}$ and study the variations of the function $f(t) := D_{KL}(X, Y)$. What well-known property do you recover?

7.7. 1. Using the fact that the map $\ln(1 + e^{-x})$ is convex, prove the following inequality (whenever the left-hand side is defined) for two arbitrary probability vectors p, q:

$$D_{KL}(p, \frac{p+q}{2}) \leq \ln \frac{2}{1 + \exp{-D_{KL}(p, q)}}.$$

2. The *Jeffrey divergence* and *Jensen–Shannon divergence* between p and q are respectively defined (see [8] for more details) by

$$D_J(p, q) := D_{KL}(p, q) + D_{KL}(q, p),$$
$$D_{JS}(p, q) := \frac{1}{2} D_{KL}(p, \frac{p+q}{2}) + \frac{1}{2} D_{KL}(q, \frac{p+q}{2}).$$

Use the previously derived inequality to prove the upper bound

$$D_{JS}(p, q) \leq \ln \frac{2}{1 + \exp{-\frac{D_{KL}(p,q)}{2}}}.$$

3. Show that it is possible to choose p and q so as to have $D_{KL}(p,q)$ arbitrarily large, whereas for all p, q we have $D_{JS}(p,q) \leq \ln 2$.

7.8. In this problem we give ourselves a (not necessarily Markovian) discrete time process (X_n), which we assume to be stationary. This means that for every choice of times n and l, the laws of the n-tuples $(X_1, \ldots, X_n)$ and $(X_{l+1}, \ldots, X_{l+n})$ are the same. The *entropy rate* of the process is the quantity $H(X)$ defined as

$$H(X) := \lim_n \frac{1}{n} H(X_1, \ldots, X_n)$$

(assuming for the moment that the limits exist; we will see under what conditions this is indeed the case). To find $H(X)$ we will make use of the following quantities:

$$\forall n \geq 1: \quad \Delta_n := H(X_{n+1}|X_n, \ldots, X_2, X_1), \quad \Delta_0 := H(X_1).$$

1. Show that the sequence (Δ_n) is nonincreasing and that it has a limit, which we shall denote by Δ.
2. Show that

$$H(X_1, \ldots, X_n) = \sum_{i=0}^{n-1} \Delta_i,$$

and deduce that $H(X) = \Delta$.
3. We now assume that the process is Markovian on a finite state space $\mathcal{S}$ (and still stationary) and that the transition matrix satisfies the assumptions of the convergence theorem for ergodic Markov chains. Let us denote the stationary distribution by π. Show that in this case, Δ is given by

$$\Delta = -\sum_x \pi(x) \sum_y p(x,y) \log p(x,y).$$

4. For the case of a two-state Markov chain, find π and write down Δ in terms of $p := p(1,1)$ and $q := p(2,2)$. For what values of p, q is this extremal? Can you draw a parallel with the result of Problem 7.1?

7.9. Consider two independent Poisson processes of intensities λ and μ, and denote by T_n^λ and T_n^μ their respective nth jump times (recall that these are gamma variables). Compute the quantity $D_{KL}(T_n^\lambda, T_n^\mu)$.

7.10. In this problem we revisit the dissipation formula (7.13) for constant-rate birth–death processes. Consider the rate matrix given by (5.7) in the case $\lambda_k = \lambda$, $\mu_k = \mu$ for all k. Assume that $\lambda < \mu$. Then as we saw in Chapter 5, the associated process has a unique detailed balance equilibrium π given by (5.11). Finally, let p_t be a solution to the Fokker–Planck equation $p_t' = p_t Q$, and set

$$H(t) := D_\phi(p_t, \pi) = \sum_y \pi(y) \phi\left(\frac{p_t(y)}{\pi(y)}\right).$$

In all that follows do all computations formally, i.e., assume always that the involved differentiation and sum operations commute.

1. Prove the relation

$$H'(t) = (-\lambda p_t(0) + \mu p_t(1))\phi'(\frac{p_t(0)}{\pi(0)})$$

$$+ \sum_{k=1}^{\infty}[\mu(p_t(k+1) - p_t(k)) - \lambda(p_t(k) - p_t(k-1))]\phi'(\frac{p_t(k)}{\pi(k)}). \quad (7.25)$$

2. Deduce

$$H'(t) = \sum_{k=0}^{\infty}(\lambda p_t(k) - \mu p_t(k+1))[\phi'(\frac{p_t(k+1)}{\pi(k+1)}) - \phi'(\frac{p_t(k)}{\pi(k)})].$$

3. Finally, using the detailed balance relations, show that

$$\frac{dH}{dt} = \sum_{k=0}^{\infty}\lambda\pi(k)[\frac{p_t(k)}{\pi(k)} - \frac{p_t(k+1)}{\pi(k+1)}][\phi'(\frac{p_t(k+1)}{\pi(k+1)}) - \phi'(\frac{p_t(k)}{\pi(k)})].$$

7.11. This is about exponential families and moment-generating functions.

1. Show that (7.19) may be obtained from (7.20) by differentiating with respect to θ and evaluating at 0.
2. Use (7.20) to recover the well-known moment-generating function of the Gaussian distribution $\mathcal{N}(\mu, \sigma^2)$:

$$\Phi(z) = \exp(\mu z + \frac{\sigma^2 z^2}{2})$$

(beware! you are dealing with vector-valued quantities).

7.12. A number of exponential families are compiled in Appendix E; pick some distributions you like, and do the corresponding algebra.

7.13. Let us examine a few entropy-maximizing distributions. We know from Theorem 7.22 that if an exponential family satisfies $k(x) = 0$, then the distribution p_θ has maximal entropy among all probability distributions of variables X satisfying $E(t(X)) = E(t(X_\theta))$. Locate in the tables of Appendix E the families for which k vanishes to see that this gives a characterization of some classical distributions as entropy maximizers. A typical example, for instance, is that among all continuous positive probability distributions having fixed expectation m, the exponential distribution of parameter $\frac{1}{m}$ has maximal entropy; can you give analogous statements for geometric, Gaussian, and Laplace variables? In each case, your statement is a particular case of Theorem 7.22; for each family, can you check by a direct computation (i.e., without using (7.21)) that the relation

$$D_{KL}(X, X_\theta) = H(X_\theta) - H(X)$$

holds?

Application: binary coding

Summary. Binary coding provides yet another example of how entropy may be used to quantify the "complexity" of things. After reviewing the basics of binary coding, we present the fundamental inequalities of Kraft and McMillan, and then in Theorem 8.9 we present the entropic bound for the length of uniquely decipherable codes. We only touch upon the aspects of coding that have to do with entropy; a more comprehensive presentation of coding theory may be found, for instance, in [24].

8.1 The basic vocabulary of binary coding

An *alphabet* is a finite set $\mathcal{A}$. The word *alphabet* is used here only to provide some intuitive support; the elements of $\mathcal{A}$ can be absolutely anything (letters, numbers, punctuation signs, vegetables), and their nature does not play any role; they will be called henceforth *symbols*. Given an alphabet $\mathcal{A}$, the set of all finite strings of elements of $\mathcal{A}$, also called *words over $\mathcal{A}$*, is denoted by $\mathcal{A}^+$. In particular, members of $\{0,1\}^+$, that is, finite strings of 0's and 1's are called *binary words*. The *length* of a word is the number of symbols it consists of. *Concatenating* two words means appending one word to another, in which case obviously the lengths add. Given two words m_1 and m_2 over the same alphabet, m_1 is said to be a *prefix* of m_2 if m_2 is obtained from m_1 by concatenating some additional (possibly empty) word.

A *binary code* c is a map from $\mathcal{A}$ into $\{0,1\}^+$, that is, a way to assign to each symbol $a \in \mathcal{A}$ a binary word $c(a)$, called its *codeword* (in some texts it is the set of codewords $c(\mathcal{A})$ that is called a *code*, and the map c is called the *encoding function*). In order to avoid confusion between $\mathcal{A}^+$ and $\{0,1\}^+$, we will henceforth reserve the terms *string* for elements of $\mathcal{A}^+$ and *codeword* for elements of $c(\mathcal{A})$.

The *code length* of the symbol a is the length of the binary word $c(a)$, which we will denote by $l_c(a)$, or simply $l(a)$ if the code c is implicitly understood. By concatenation, the map c is extended to the set of words over $\mathcal{A}$, to give a map (called the *extension of c* and denoted by c^+ or sometimes simply c) from

J.-F. Collet, *Discrete Stochastic Processes and Applications,*
Universitext, https://doi.org/10.1007/978-3-319-74018-8_8

$\mathcal{A}^+$ into $\{0,1\}^+$ (but beware, *codeword* means image of a symbol only). The first requirement for a code is that it should be possible to recover a symbol from its codeword, or more generally, a string from the corresponding binary word; let us make this precise by giving a clean definition:

Definition 8.1. *A code c is said to be nonsingular if no two different symbols may be mapped to the same codeword, and uniquely decipherable if no two different strings may be mapped to the same binary word.*

In more mathematical parlance, a code c is nonsingular if and only if the map c is injective, and it is uniquely decipherable if and only if the map c^+ is injective (or equivalently, if c^+ is nonsingular). Obviously, unique decipherability implies nonsingularity, but it is easy to convince oneself that the two are not equivalent. For instance, over a three-symbol alphabet x, y, z, the code c defined by

$$c(x) = 0, \ c(y) = 1, \ c(z) = 01$$

is clearly nonsingular, but it is not uniquely decipherable, since the words xy and z are both coded by the same word 01. If one wishes to avoid this type of situation, the following condition emerges:

Definition 8.2. *A code c is said to be instantaneous, or to be a prefix code (whereas logically it should have been called a no-prefix code) if no codeword is a prefix of another codeword.*

The adjective *instantaneous* expresses the fact that if a long message is read from left to right, each codeword will be instantaneously decoded: once a codeword $c(x)$ has been recognized, we know it cannot be the beginning of some longer codeword, so we can move to the next. Instantaneity clearly implies unique decipherability: if two strings $x_1 \cdots x_n$ and $y_1 \cdots y_p$ are mapped to the same binary word, then $c(x_1)$ must be a prefix of $c(y_1)$ if $l_c(x_1) \leq l_c(y_1)$, and if $l_c(x_1) \geq l_c(y_1)$, then $c(y_1)$ must be a prefix of $c(x_1)$. In either case we have $x_1 = y_1$, and continuing with the other symbols, we see that both strings must be equal. The converse is, however, false: there are uniquely decipherable codes that are not instantaneous. For instance, over a three-symbol alphabet, the set of codewords $\{1, 10, 100\}$ gives a uniquely decipherable code: if in a sequence of codewords one introduces a vertical separator before each 1, it is easy to see that this gives the decoding map.

To summarize, we have three classes of codes, each of which is included in the next one: instantaneous codes, uniquely decipherable codes, and nonsingular codes. If we work with variable lengths, both inclusions are strict, and if we restrict ourselves to fixed-length codes (that is, if all codewords are to have the same length), then it is easy to check that all three classes coincide.

Let us close with a remark. The instantaneity condition involves only the codewords and does not say anything about the map c^+. It is obvious that the extension c^+ of an instantaneous code cannot be instantaneous on $\mathcal{A}$: if a string m_1 is a prefix of another string m_2, then $c(m_1)$ is a prefix of $c(m_2)$. As a matter of fact, this turns out to be the only way to generate prefixes for c^+. More precisely, by adapting the reasoning above to show that instantaneity implies unique decipherability, you will have no trouble showing the following:

Theorem 8.3. *Let c be instantaneous. If two strings $m_1, m_2 \in \mathcal{A}^+$ are such that $c(m_1)$ is a prefix of $c(m_2)$, then m_1 is a prefix of m_2.*

8.2 Instantaneous codes and binary trees

The representation of an instantaneous code as a binary tree is interesting from two points of view, first because it gives a quick proof of the Kraft–MacMillan inequality (see below), which will be useful for estimating the average length of codes, and second (as we shall see about the Huffman algorithm), because some codes are obtained precisely by constructing a binary tree.

First some vocabulary from discrete mathematics:

A *binary tree* is a rooted tree in which each node has at most two children, and the nodes that have no children are called the *leaves* of the tree (let us recall that mathematicians, just like computer scientists, have never set foot in a forest, so that they draw trees upside down, with the root at the top and the leaves at the bottom). A binary tree is said to be *complete* if each node that is not a leaf has exactly two children, which we will call the *left child* and *right child*. We will stick to the convention according to which the root of the tree is assigned depth 0, and the *depth* of any other node is the number of edges that connect it to the root. The *depth* of the tree is then defined to be the maximum of all depths of nodes (which is therefore attained at some leaf). For any given depth there is therefore a unique complete binary tree; every binary tree of the same depth may be constructed from this unique complete tree by selecting a certain number $n \geq 0$ nodes and erasing all their descendants.

Choosing a binary code roughly amounts to choosing a certain number of nodes and assigning to each of them a binary string whose length is equal to the depth of the node; in order to do this, we can number the nodes according to the lexicographic order of the alphabet $\{0, 1\}$. If you are not familiar with this, we strongly recommend that you look it up in any basic textbook on data structures. However, in the present case it is sufficient to know that this simply means assigning 0 to the left child of the root, 1 to its right child, and then working your way down by repeating the process. Here, for instance, is the result for depth 4:

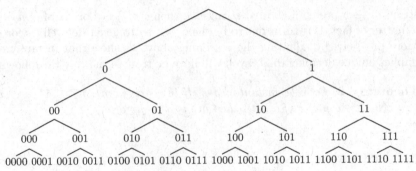

From this complete tree one obtains an instantaneous code by selecting a certain number of nodes and then erasing for each of those nodes the corresponding subtree, as, for example, in the following:

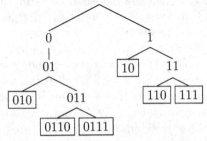

The codewords are then simply the labels carried by the leaves.

Conversely, given an instantaneous code, if we denote by l the maximum length of all its codewords, we may visualize the code as the tree obtained from the complete binary tree of depth l by suppressing the subtrees associated with the codewords; we see that another code having the same set of codewords (but assigned to different symbols) would correspond to the same tree; in other words, it is not exactly the map c that is encoded in the tree, but rather the set $c(\mathcal{A})$ of its codewords.

8.3 The Kraft–McMillan inequality

8.3.1 The instantaneous case

The fundamental inequality used for estimating the length of instantaneous codes is Kraft's inequality, which is a statement about binary trees as well as about instantaneous codes (this should come as no surprise given the correspondence we have described in the preceding section). Let us begin with the version for trees:

Theorem 8.4 (Kraft's inequality for binary trees). *Let $\mathcal{F}$ denote the set of leaves of a binary tree $\mathcal{T}$, and for each leaf x let $d(x)$ be its depth. Then*

$$\sum_{x \in \mathcal{F}} 2^{-d(x)} \leq 1. \tag{8.1}$$

Proof. Let l be the depth of $\mathcal{T}$.

If all leaves have depth l then (8.1) just says that the number of leaves is less than or equal to 2^l. If some leaves x satisfy $d(x) < l$, then every such leaf has in the complete binary tree of depth l a number of descendants that is equal to $2^{l-d(x)}$, and those are missing in $\mathcal{T}$. Summing over all leaves of $\mathcal{T}$, we obtain the number of forbidden positions at depth l, which is necessarily less than the total number of positions at depth l, i.e., 2^l. This means that

$$\sum_{x \in \mathcal{F}} 2^{l-d(x)} \leq 2^l,$$

from which (8.1) follows. $\square$

Using the correspondence between trees and instantaneous codes, we immediately obtain the version for codes:

Theorem 8.5 (Kraft's inequality for instantaneous codes). *Let c be an instantaneous code over an alphabet $\mathcal{A}$, and for each symbol $x \in \mathcal{A}$ let $l_c(x)$ be the length of the corresponding codeword. Then the following inequality holds:*

$$\sum_{x \in \mathcal{A}} 2^{-l_c(x)} \leq 1. \tag{8.2}$$

It turns out that the converse is valid, in such a way that Kraft's inequality provides a characterization of instantaneous codes:

Theorem 8.6. *Let $\mathcal{A} := \{x_1, \ldots, x_p\}$ be an alphabet with p elements, and let $l_1 \leq l_2 \cdots \leq l_p$ be p numbers satisfying the inequality*

$$\sum_{i=1}^{p} 2^{-l_i} \leq 1.$$

Then there exists an instantaneous code c over $\mathcal{A}$ whose codewords have lengths l_i, i.e., such that

$$l_c(x_i) = l_i, \quad 1 \leq i \leq p.$$

Proof. Draw the complete binary tree of depth l_m, and label it lexicographically as described previously. It then suffices to choose any node at depth l_1 and then erase its subtree (which means all its descendants) and to keep going with this pruned tree with a node at depth l_2, and so forth. As in the proof of the Kraft inequality for trees, the assumed inequality guarantees that there will be enough room in the tree to complete the construction and reach depth l_m. $\square$

As a curiosity, let us close with a probabilistic proof of Kraft's inequality for instantaneous codes that (almost) does not rely on the representation as a tree: Let c be an instantaneous code, its codewords having maximal length l_m. Let us give ourselves a family of i.i.d. Bernoulli variables of parameter

$\frac{1}{2}$ (in other words, we shall draw a sequence of l_m digits, all of them drawn uniformly and without memory).

For each symbol $x \in \mathcal{A}$ let us denote by E_x the event "$c(x)$ is a prefix of the obtained sequence." In other words, E_x is realized if and only if in the complete binary tree of depth l_m, the node corresponding to $c(x)$ is an ancestor of the obtained sequence; since this one was drawn uniformly and independently, we have

$$P(E_x) = \frac{2^{l_m - l(x)}}{2^{l_m}} = 2^{-l(x)}.$$

Now, c being instantaneous, these events are disjoint, and therefore,

$$\sum_{x \in \mathcal{A}} 2^{-l(x)} = \sum_{x \in \mathcal{A}} P(E_x) = P(\bigcup_{x \in \mathcal{A}} E_x) \le 1.$$

□

8.3.2 The uniquely decipherable case

In fact, the Kraft inequality still holds for the larger class of uniquely decipherable codes, in which case it changes its name to become the *Kraft–McMillan inequality*. Of course, this generalizes the previous inequalities, but here we present it separately, because the extension of the result to a much larger class requires a significantly new (and not so easy) argument:

Theorem 8.7 (Kraft–McMillan inequality). *Every uniquely decipherable code satisfies* (8.2).

Proof. Let m be the cardinal of $\{\mathcal{A}\}$, and let $l_1 \le \cdots \le l_m$ be the codeword lengths arranged in increasing order.

Calling S the sum we wish to bound, the trick is to study the growth of S^n as n increases:

$$S^n = [\sum_{x \in \mathcal{A}} 2^{-l(x)}]^n = \sum_{i_1=1}^{m} \cdots \sum_{i_n=1}^{m} 2^{-(l_{i_1} + \cdots + l_{i_n})}.$$

Choosing an n-tuple $(i_1 \cdots i_n)$ amounts to choosing an n-symbol string over $\{\mathcal{A}\}$; the exponent in the last expression represents the length of the coding of this string, which is therefore less than nl_m. The multiple sum above runs over all n-symbol strings, and by partitioning it according to coding length, we obtain

$$S^n = \sum_{k=1}^{nl_m} N_k 2^{-k},$$

where $N_k \ge 0$ denotes the number of n-symbol strings whose coding has length k. Since c is uniquely decipherable, this number is necessarily less than the total number of possible codewords of length k, i.e., 2^k. This yields the bound

$$S^n \leq n l_m,$$

which implies $S \leq 1$ (surely you must know that if $|z| > 1$, then $\frac{|z|^n}{n} \to \infty$ as $n \to \infty$). $\square$

8.4 Code length and entropy

Kraft's inequality tells us precisely in what way the decipherability constraint imposes long codewords; if one wishes to transmit data while minimizing volumes, a second problem that now arises is how to cleverly distribute these codeword lengths. Therefore, we need to have a precise notion of "rare" symbols as opposed to "common" ones.

We shall call a random variable with values in $\mathcal{A}$ a *source over* $\mathcal{A}$, and we will denote by p the corresponding vector: $p(x) := P(X = x)$. (Note that in this notation X is implicit, since we never use two different sources at a time; for practical purposes $p(x)$ will be approximated by the frequency of the symbol x.)

The basic idea now is that if we need to have some large codeword lengths as Kraft's inequality requires, these should go to those symbols x for which $p(x)$ is large. The average length that will be transmitted for one symbol at a time will then be the expectation of $l_c(X)$, which we call the *average code length of c*:

Definition 8.8. *The average code length of c is the quantity*

$$L(c) := \sum_{x \in \mathcal{A}} l_c(x) p(x).$$

Now comes the main result, which shows that this average code length cannot be better than the entropy of the source:

Theorem 8.9. *Every uniquely decipherable code c satisfies*

$$L(c) \geq H_2(X).$$

In the statement above, the subscript 2 indicates that we are using base-2 logarithms.

Proof. We begin by computing the difference

$$L(c) - H_2(X) = \sum_{x \in \mathcal{A}} l_c(x) p(x) + \sum_{x \in \mathcal{A}} l_c(x) \log_2 p(x) = \sum_{x \in \mathcal{A}} p(x) \log_2 \frac{p(x)}{2^{-l_c(x)}}.$$

This last sum looks like a Kullback–Leibler divergence, but it is not exactly one, because the denominators are not the components of a probability vector; this leads us to normalize them by setting

$$q(x) := \frac{2^{-l_c(x)}}{\sum_y 2^{-l_c(y)}}.$$

This defines a probability vector q, and now we have

$$L(c) - H_2(X) = D_{KL}(p, q) - \log_2(\sum_y 2^{-l_c(y)}).$$

From the Kraft–McMillan inequality (8.2), the logarithm is nonpositive, which establishes the inequality. □

Of course, the next question is whether it is possible to approach this bound. The answer is yes, within one unit:

Theorem 8.10. *There exists an instantaneous code c satisfying*

$$L(c) \leq H_2(X) + 1.$$

Proof. Let us first express $H(X) + 1$ as a unique sum:

$$H(X) + 1 = \sum_{x \in \mathcal{A}} p(x)(1 - \log_2 p(x)).$$

We wish to find a code c with average code length close to this expression; this would indeed be an average code length if the quantities $1 - \log_2 p(x)$ were integers; hence the idea of considering its integer part and setting

$$l_x := 1 + [-\log_2 p(x)],$$

where $[\cdot]$ means the integer part. Recall that this is uniquely defined by the inequalities

$$\forall z \geq 0 : z - 1 < [z] \leq z.$$

We can therefore bound the length as follows:

$$l_x \geq -\log_2 p(x),$$

whence

$$\sum_x 2^{-l_x} \leq \sum_{x \in \mathcal{A}} p(x) = 1.$$

We may now invoke Theorem 8.6 to obtain an instantaneous code c satisfying $l_c(x) = l_x$ for all x, and we may bound its average code length:

$$L(c) = \sum_{x \in \mathcal{A}} p(x)(1 + [-\log_2 p(x)]) \leq \sum_{x \in \mathcal{A}} p(x)(1 - \log_2 p(x)) = H(X) + 1.$$

□

8.5 Problems

8.1. This problem is about entropy of a binary sequence.

1. Let X be a Bernoulli variable and put $p := P(X = 1)$. Compute its entropy $H(X)$.
2. We now give ourselves a source emitting digits independently, each one according to X. Let S be a given sequence of n digits, and let k be the number of 1's in S. What is the probability $P(S)$ of emission of that particular sequence?
3. We now consider the limit $n \to \infty$, with k close to np. Find in this asymptotic an equivalent of the quantity $\log P(S)$ in terms of $H(X)$.

8.2. Consider a source emitting a symbol from a four-symbol alphabet, the emission probabilities being $\frac{1}{2}, \frac{1}{4}, \frac{1}{8}, \frac{1}{8}$. Show that the choice of codewords $0, 10, 110, 111$ gives an optimal code.

8.3. Let X be a source with emission probabilities $(p_i)_i$. The set of all negative powers of 2 provides a grid on the interval $[0, 1]$, which means that every p_i lies between two of them:

$$2^{-l_i} \leq p_i \leq 2^{-l_i+1},$$

with $l_i \in \mathbb{N}$.

1. Show that if a code uses the set of lengths $(l_i)_i$, then it satisfies

$$H(X) \leq L(c) \leq H(X) + 1.$$

2. Consider the emission probabilities

$$p_1 = \frac{1}{2}, \ p_2 = \frac{1}{4}, \ p_3 = \frac{1}{8}, \ p_4 = \frac{1}{16}, \ p_5 = \frac{1}{16},$$

and corresponding codewords

$$c_1 = 1, \ c_2 = 10, \ c_3 = 110, \ c_4 = 1110, \ c_5 = 1111.$$

Compute the entropy $H(X)$ and the code length $L(c)$. Your conclusion?

8.4. In this problem we are going to see that coding in blocks reduces the average length per symbol. Let X be a source. We define a new random variable Y by taking a certain number s of independent copies of X, i.e., we set $Y = (X_1, X_2, \ldots, X_s)$, where the X_i are i.i.d. and distributed as X.

1. Show that $H(Y) = sH(X)$.
2. Deduce that it is possible to find a code c for the source Y satisfying

$$H(X) \leq \frac{L(c)}{s} \leq H(X) + \frac{1}{s}.$$

3. What interpretation can you find for the quantity $\frac{L(c)}{s}$?

8.5. If $x, y \in \{0,1\}^q$ are two binary strings of length q, their *Hamming distance* is the integer defined by

$$d_H(x,y) := \sum_{i=1}^{q} |x_i - y_i|.$$

In other words, $d_H(x,y)$ is the number of digits in which x and y differ. Show that this is indeed a distance on $\{0,1\}^q$.

8.6. Imagine that binary strings of length q are transmitted through a noisy channel, so that some bits will be incorrectly transmitted. We assume the channel to be *symmetric*, which means that the probability r of a transmission error for a digit is the same whether this digit is a 0 or a 1, and we assume that all bits are transmitted independently. Mathematically, this means that if we denote by E_i the Bernoulli variable that is equal to 1 if bit number i is transmitted incorrectly, we have a collection of q i.i.d. Bernoulli variables $E_1, \ldots, E_q$ of parameter r.

1. For a given (nonrandom) original string $x \in \{0,1\}^q$, what is the distribution of the number of errors in its transmission?
2. For two given strings $x, y \in \{0,1\}^q$, show that the probability that x is transmitted as y is given by

$$P(x \to y) = (1-r)^n \left(\frac{r}{1-r}\right)^{d_H(x,y)}.$$

3. If y was obtained from the noisy channel and we wonder which codeword $x \in c(\mathcal{A})$ was transmitted, one might think that x should be a codeword as close as possible to y in the Hamming distance. Thus we are led to wonder whether for fixed y, the quantity $P(x \to y)$ is a decreasing function of $d_H(x,y)$. For what values of r is this the case? Can you see what goes wrong if your condition on r is not satisfied?

Some useful facts from calculus

A.1 Croft's lemma

Croft's lemma comes in handy (in a slightly generalized version) when one is trying to establish regularity of the semigroup of a continuous-time Markov process by the use of skeletons; the usual statement is the following:

Lemma A.1 (Croft's lemma). *Let $f : (A, \infty) \to \mathbb{C}$ be a continuous map for some $A \geq 0$ such that for every $t > A$, we have*

$$\lim_{n \to \infty} f(nt) = l,$$

where l is a real number. Then

$$\lim_{x \to \infty} f(x) = l.$$

In Chapter 4 we use a version of this result near 0:

Lemma A.2 (Croft's lemma near 0). *Let $v : (0, B) \to \mathbb{C}$ be a continuous map for some $B \leq \infty$ such that for every $t \in (0, B)$, we have*

$$\lim_{n \to \infty} v(\frac{t}{n}) = l,$$

where l is a real number. Then

$$\lim_{h \to 0^+} v(h) = l.$$

Both versions are clearly equivalent (consider $v(t) := f(\frac{1}{t})$); let us prove the second one:

Proof. Of course (considering $v - l$), we may take $l = 0$. For fixed $\epsilon > 0$ and $p \in \mathbb{N}$, consider the set

$$C_p := \{t \in (0, l) : \forall n \geq p : |v(\frac{t}{n})| \leq \epsilon\}.$$

© Springer International Publishing AG, part of Springer Nature 2018
J.-F. Collet, *Discrete Stochastic Processes and Applications*,
Universitext, https://doi.org/10.1007/978-3-319-74018-8

A Some useful facts from calculus

The sets $(C_p)_{p\in\mathbb{N}}$ form an increasing sequence, and since v is continuous on the interval $(0,l)$, they are all closed. On the other hand, for all $t \in (0,l)$, the convergence of $v(\frac{t}{n})$ tells us that $t \in C_p$ for some p large enough, and therefore, we have

$$(0,l) = \bigcup_{p\geq 1} C_p.$$

Baire's theorem (see, for instance, [32]) now tells us that at least one set C_p must have nonempty interior, i.e., must contain an interval (a,b). This means that for all $t \in (a,b)$ and $n \geq p$ we have $|v(\frac{t}{n})| \leq \epsilon$, or in other words, for all $x \in (\frac{a}{n}, \frac{b}{n})$ we have $|v(x)| \leq \epsilon$. Summarizing, we have shown the following:

$$\forall x \in \bigcup_{n\geq p} (\frac{a}{n}, \frac{b}{n}): \quad |v(x)| \leq \epsilon.$$

Consider now the sequence of intervals $I_n := (\frac{a}{n}, \frac{b}{n})$ as n increases. The left endpoint $\frac{a}{n}$ clearly converges to 0, and these intervals eventually overlap. More precisely, the intervals I_n and I_{n+1} overlap if and only if $\frac{b}{n+1} > \frac{a}{n}$, or equivalently if and only if $a < (1 + \frac{1}{n})b$. Since $a < b$, this is satisfied for all n large enough, say $n \geq n_0$. We may take $n_0 \geq p$, thus guaranteeing that the union $\cup_{n\geq n_0} I_n$ contains the interval $(0, \frac{a}{n_0})$. Setting $x_0 := \frac{a}{n_0}$, we get $|v(x)| \leq \epsilon$ for all $x \in (0, x_0)$, which shows the desired convergence to 0. $\square$

When discretizing a continuous-time Markov chain in Chapter 4 we make use of the following generalization:

Lemma A.3. *Let $v : (0, B) \to \mathbb{C}$ be a continuous map for some $B \leq \infty$ such that for all $t \in (0, B)$ we have*

$$\lim_{n\to\infty} v(\frac{t}{n}) = \alpha(t),$$

where α is monotonic, right continuous, or left continuous. Then α has to be constant: $\alpha(t) = l$ for all $t \in (0, B)$, and

$$\lim_{h\to 0+} v(h) = l.$$

Proof. In view of Lemma A.2, we need to show only that α is constant; let us first show that for every rational number $r \in (0, 1]$ and every number $s \in (0, B)$, we have $\alpha(rs) = \alpha(s)$. Writing $r = \frac{k}{q}$, we have

$$v(r\frac{s}{kn}) \to \alpha(rs)$$

(consider the sequence (kn) and $t = rs$), and on the other hand,

$$v(r\frac{s}{kn}) = v(\frac{s}{qn}) \to \alpha(s),$$

(consider the sequence (qn)), from which we get $\alpha(rs) = \alpha(s)$.

Now let $t_1, t_2 \in (0, B)$ be any two numbers with $t_1 < t_2$. We may use a sequence of rational numbers (r_n) decreasing to $\frac{t_1}{t_2}$ and a sequence of rational numbers (s_n) decreasing to $\frac{t_1}{t_2}$, with $r_n, s_n \leq 1$. If α is nonincreasing, we have

$$\alpha(t_2) = \alpha(s_n t_2) \leq \alpha(t_1) \leq \alpha(r_n t_2) = \alpha(t_2),$$

and therefore $\alpha(t_1) = \alpha(t_2)$; obviously, this argument (with reversed inequalities) is also valid if α is nondecreasing. If α is right continuous, then letting $n \to \infty$ in the upper bound, we see that $\alpha(t_1) \leq \alpha(t_2)$; therefore α is nondecreasing, and we are done. Similarly, if α is left continuous, we conclude that it has to be nonincreasing. $\square$

A.2 A characterization of exponential functions and distributions

Let us begin with the following characterization of exponential functions, which we shall need in Chapter 4:

Lemma A.4. *Assume that u and α are two maps defined on the interval $[0, \infty) \to \mathbb{R}$ such that*

$$\forall t > 0 : \quad \lim_{n \to \infty} u(\frac{t}{n})^n = \alpha(t).$$

Assume, moreover, that the map α is monotonic, right continuous, or left continuous. Then α is given by

$$\forall t > 0 : \quad \alpha(t) = \alpha(1)^t. \tag{A.1}$$

Proof. We are first going to show that

$$\forall t > 0, \forall k \in \mathbb{N} : \quad \alpha(kt) = \alpha(t)^k. \tag{A.2}$$

Let $k \geq 1$ be a fixed integer; considering the extracted sequence $n_k := kn$, as $n \to \infty$ we obtain

$$u(\frac{t}{nk})^{nk} \to \alpha(t).$$

On the other hand, we have

$$u(\frac{t}{nk})^n \to \alpha(\frac{t}{k}),$$

and therefore,

$$\alpha(\frac{t}{k})^k = \alpha(t),$$

which on changing t into kt yields (A.2).

A Some useful facts from calculus

Let us first prove (A.1) for rational numbers: for nonzero $q \in \mathbb{N}$, taking $t = \frac{1}{q}$ and $k = q$, we have $\alpha(\frac{1}{q}) = \alpha(1)^{\frac{1}{q}}$. Thus for all $k, q \in \mathbb{N}$ with $q \neq 0$, we obtain

$$\alpha(\frac{k}{q}) = \alpha(\frac{1}{q})^k = \alpha(1)^{\frac{k}{q}}.$$

We may now use monotonicity to move on to real numbers: for $t > 0$, let r_k be a sequence of rational numbers decreasing to t, and s_k a sequence of rational numbers increasing to t. If α is (for instance) increasing, then for all k we have

$$\alpha(s_k) = \alpha(1)^{s_k} \leq \alpha(t) \leq \alpha(r_k) = \alpha(1)^{r_k}.$$

Taking the limit $k \to \infty$ yields $\alpha(t) = \alpha(1)^t$. If α is left continuous, it suffices to use the sequence s, and the sequence r if α is right continuous. $\square$

A byproduct of the proof is that every function α satisfying (A.2) must be of the form (A.1); this can be used to prove the following characterization of exponential and geometric distributions (commonly referred to as the *memoryless property*):

Corollary A.5. *1. A positive absolutely continuous random variable X satisfies the relation*
$$\forall t, s \geq 0: \quad P(X > t + s) = P(X > t)P(X > s) \qquad (A.3)$$

if and only if it follows an exponential distribution.

2. A positive random variable X taking values in $\mathbb{N}$ satisfies (A.3) if and only if it follows a geometric distribution.

Proof. In either case, note that $P(X > t)$ is a nonincreasing function of t. If X is exponential and λ is its parameter, you may check that

$$P(X > t) = e^{-\lambda t},$$

which implies (A.3); for necessity, note that (A.3) implies that $\alpha(t) := P(X > t)$ satisfies (A.2); therefore, if we put $\lambda := -\ln P(X > 1)$, then Lemma A.4 gives us

$$P(X < t) = 1 - e^{-\lambda t}.$$

Thus by differentiating, we obtain $f_X(t) = \lambda e^{-\lambda t}$. For the discrete case, if X is geometric of parameter p, you may check that

$$\forall t > 0: \quad P(X > t) = (1 - p)^t,$$

and thus (A.3) is satisfied. Conversely, if (A.3) is true, then by setting $p := P(X = 1)$, Lemma A.4 gives us

$$\forall t \geq 1: \quad P(X > t) = (1 - p)^t.$$

We may now obtain the probability mass function by taking a difference:

$$\forall t \geq 1: \quad P(X = t) = P(X > t - 1) - P(X > t) = p(1 - p)^{t-1}.$$

Therefore, X is geometric. $\square$

Another consequence of Lemma A.4 is the well-known result that if a continuous function f satisfies the so-called functional equation for the exponential,

$$f(t + s) = f(t)f(s), \tag{A.4}$$

then it is an exponential function; if in (A.4) we replace equality by an inequality, we have a comparison result, which we shall use in Chapter 4:

Lemma A.6. *Assume that f is a map defined on some interval $[0, T)$ with $T \leq \infty$, that it has a right derivative at 0, and that it satisfies the following inequality:*

$$\forall t, s / 0 \leq t + s < T : \quad f(t + s) \geq f(t)f(s).$$

Assume, moreover, that $f(0) = 1$. Then

$$\forall t \geq 0 : \quad f(t) \geq \exp\left(tf'(0)\right) \geq 1 + tf'(0). \tag{A.5}$$

Note that since $\exp\left(tf'(0)\right)$ is the solution to the functional equation (A.4), this says that every function satisfying the corresponding inequality is necessarily larger than this solution, whence the name *comparison* result.

Proof. For all $t \in [0, T)$ and $n \in \mathbb{N}$ we have $f(t) \geq f(\frac{t}{n})^n$; since f is right continuous at 0, it follows that for fixed t, if n is large enough, then $f(\frac{t}{n})$ is close to 1, so we may manipulate logarithms without a second thought; then as $n \to \infty$ we have

$$\ln f(t) \geq n \ln f(\frac{t}{n}) \sim n(f(\frac{t}{n}) - 1) = t\frac{f(\frac{t}{n}) - 1}{\frac{t}{n}} \to tf'(0).$$

This proves the first half of inequality (A.5), and the second half follows from the classical inequality $e^h \geq 1 + h$ for all $h \geq 0$. $\quad\square$

A.3 Countable sums

If $(z_i)_{n \in \mathbb{N}}$ is a sequence of complex numbers, we all know what is meant by the sum of the associated series (whenever the limit exists):

$$\sum_{n=0}^{\infty} z_n := \lim_{N \to \infty} \sum_{n=0}^{N} z_n.$$

If, on the other hand, we have a countable set $\mathcal{S}$ of complex numbers, how can we define the sum $\sum_{z \in \mathcal{S}} z$ of all its elements? The surprising fact is that if we define a numbering of the elements of $\mathcal{S}$, i.e., if we choose a one-to-one map $z : i \in \mathbb{N} \mapsto z_i \in \mathcal{S}$ (after all, countable sets are exactly those for which such a map may be found), then the sum $\sum_{i=0}^{\infty} z_i$ depends on the choice of the map z. More precisely, if $\pi : \mathbb{N} \to \mathbb{N}$ is a bijective map and $(z_i)_{n \in \mathbb{N}}$ is a sequence of complex numbers, in general the two following sums will differ:

A Some useful facts from calculus

$$\sum_{i=0}^{\infty} z_i \neq \sum_{j=0}^{\infty} z_{\pi(j)}$$

(you should play around with the sequence $(-1)^n$ to convince yourself of this). In other words, the value we get for the sum depends on the order in which we throw in the elements of S. The good news is that this does not happen when we deal with nonnegative numbers:

Theorem A.7. *Let $(z_i)_{n \in \mathbb{N}}$ be a sequence of nonnegative numbers, $S :=$ $\{z_i, \ i \in \mathbb{N}\}$ the set of all its terms, and $\mathcal{F}$ the collection of all finite subsets of S. Define the sum of all elements in S as*

$$\sum_{x \in S} x := \sup\{\sum_{x \in F} x, \quad F \in \mathcal{F}\}.$$

Then

$$\sum_{x \in S} x = \sum_{i=0}^{\infty} z_i$$

(whether the series is finite or infinite).

Proof. Every partial sum is obviously less than $\sum_{x \in S}$; on the other hand, if F is finite, then for n large enough we have $F \subset \{1, \ldots, \}$, so that

$$\sum_{x \in F} x \leq \sum_{i=0}^{n} z_i \leq \sum_{i=0}^{\infty} z_i.$$

$\square$

The sum $\sum_{x \in S}$ as we just defined it depends only on the set S; therefore, any other numbering of its elements would give the same value for the sum of the series.

In Chapter 4 we often encounter functions defined as sums of series; the main tool for establishing continuity of such sums is the following result, which is a special case of the Lebesgue dominated convergence theorem:

Theorem A.8. *Let $(u_n)_{n \in \mathbb{N}}$ be a sequence of maps defined on some interval $[0, T)$ with $T \leq \infty$ such that each u_n is right continuous at 0 that satisfies the following "domination" inequality:*

$$\forall t \in [0, T), \forall n \in \mathbb{N}: \quad |u_n(t)| \leq v_n,$$

where $(v_n)_{n \in \mathbb{N}}$ is a sequence of nonnegative numbers such that the series $\sum v_n$ is finite. Then the sum $\sum_n u_n(t)$ is right continuous at 0:

$$\lim_{t \to 0^+} \sum_{n=0}^{\infty} u_n(t) = \sum_{n=0}^{\infty} u_n(0).$$

176

The proof is a very easy $\frac{\epsilon}{2} + \frac{\epsilon}{2}$ exercise, which we leave to the reader. More interestingly, the essence of this result is that the tail of the infinite sum is controlled by the domination inequality, which ensures that the infinite sum behaves exactly as a finite one would. Also, note that we formulated our result here as a statement on right continuity at 0, but it may be easily adapted to prove (two-sided) continuity of the sum of a series of continuous functions at a given point.

A.4 Right continuous and right constant functions

For a set T, the notion of continuous maps with values in T depends on the topology we use. Recall that the *discrete topology* on T is the one for which all sets are open. This means that all one-point sets $\{x\}$ are open, and as a consequence, every convergent sequence of points in T has to be constant after a certain point. In other words, if $t_n \in T \to t \in T$, then for some n_0 we have $t_n = t_{n_0}$ for all $n \geq n_0$. It turns out that the right-constancy condition we saw in Chapter 4 is exactly equivalent to right continuity for this discrete topology:

Lemma A.9. *Let T be any set endowed with the discrete topology and I an open interval of $\mathbb{R}$. A map $f : I \to T$ is right continuous if and only if it is right constant, meaning that for every $x \in I$, there exists some $h > 0$ such that $x + h \in I$, and f is constant on the interval $[x, x + h]$.*

Proof. Trivially, right constancy implies right continuity; to prove the converse, let us argue by contradiction, i.e., assume that f is right continuous and no h satisfies the above condition, and consider a sequence $h_n \to 0$. Then we may find a subsequence (call it still h_n) for which each term $f(x + h_n)$ differs from the preceding one $f(x + h_{n-1})$. On the other hand, since f is right continuous at x, we must have $f(x + h_n) \to f(x)$, which implies that the sequence $f(x + h_n)$ has to be eventually constant, a contradiction. $\square$

To put it in more graphical terms: take any point on the graph of f; then to the right of this point, the graph of x will locally be a horizontal segment. Note that this result says nothing about the width of this step: it may be very small, and we may even have situations in which the width of consecutive steps forms a sequence going to 0; a typical example is given by

$$f(x) := n \quad \forall x \in [\frac{1}{2^n}, \frac{1}{2^n} + \frac{1}{2^{n+1}}), \forall n \in \mathbb{N}.$$

Obviously, right constancy will imply right continuity whatever topology we use on T; however, the converse implication is not true in general. Consider, for instance, the countable set

$$T := \{0\} \cup \{\frac{1}{n}, n \geq 1\}$$

A Some useful facts from calculus

endowed with the absolute-value distance (inherited from $\mathbb{R}$). Define the map $f : [0, \infty) \to T$ by

$$\forall x \in [\frac{1}{2^{n+1}}, \frac{1}{2^n}) : \quad f(x) = \frac{1}{n},$$

and $f(0) = 0$. Then you may check as an exercise that f is right continuous on $[0, \infty)$, although it does not satisfy the right constancy property at 0. Finally, note that there is no contradiction with Lemma A.9: the sequence $\frac{1}{n}$ does not converge to 0 in the discrete topology on T.

A.5 The Gamma function

The Euler Gamma function is defined as follows:

$$\forall z > 0 : \quad \Gamma(z) := \int_0^\infty t^{z-1} e^{-t} \, dt. \tag{A.6}$$

In fact, this definition makes sense for every complex number z with $\Re z > 0$, but in this text we have no use for that. As is well known, this is a generalization of the factorial in the sense that for every positive integer n we have $\Gamma(n) = (n-1)!$; the asymptotics for z large are provided by the famous Stirling formula:

$$z \to \infty : \quad \Gamma(z+1) \sim \sqrt{2\pi z}(\frac{z}{e})^z. \tag{A.7}$$

Another important property of the Gamma function that we will need is its log-convexity (see Problem 6.12 for a proof, and the monograph [1] for other properties of the Γ function). This means that for every set of n nonnegative numbers $\lambda_1, \ldots, \lambda_n$ satisfying $\lambda_1 + \cdots + \lambda_n = 1$ and every set of nonnegative numbers $x_1, \ldots, x_n$, we have

$$\ln \Gamma(\sum \lambda_i x_i) \le \sum \lambda_i \ln \Gamma(x_i).$$

As a particular case, if we take $X_i = n_i + 1$ (where n_i is a nonnegative integer) and $\lambda_i = \frac{1}{n}$, we obtain the following inequality, which we shall make use of in the study of random walks:

$$\prod_{i=1}^n n_i! \ge [\Gamma(1 + \sum_{i=1}^n \frac{n_i}{d})]^n. \tag{A.8}$$

A.6 The Laplace transform

This is a minimal reminder of a few properties of the Laplace transform that we make use of in the text; as usual, we take a user-oriented approach and do not seek general or sharp results (for an excellent introduction to this topic we recommend taking a look at [33]). Recall that if $f : [0, \infty) \to \mathbb{R}$ is a function, its Laplace transform is the map Lf defined by

$$(Lf)(z) := \int_0^\infty e^{-zt} f(t) \, dt$$

(it may be checked that $(Lf)(z)$ as defined here makes sense for every complex number with real part greater than some number α that depends on f; but here we shall focus on only the algebraic properties of the Laplace transform without getting into summability issues). The first elementary property of the Laplace transform is its behavior with respect to differentiation, which is what makes it so useful as a tool for constant-coefficient differential equations. More precisely, if f is differentiable, by a very easy integration by parts you may show that

$$(Lf')(z) = z(Lf)(z) - f(0).$$

Another key property is its behavior with respect to the convolution product. Recall that if f and g are two maps defined on $[0, \infty)$, we may extend them to $\mathbb{R}$ by adding the value 0, i.e., replacing them by $f\mathbb{1}_{[0,\infty)}$ and $g\mathbb{1}_{[0,\infty)}$ respectively. Then their convolution product $f * g$ which is defined by

$$(f * g)(t) := \int_\mathbb{R} f(t - s)g(s) \, ds$$

reduces to an integral on a bounded interval:

$$(f * g)(t) = \int_0^t f(t - s)g(s) \, ds.$$

The Laplace transform turns convolution products into usual products: for any two maps f, g and $t > 0$ we have

$$(L(f * g))(t) = Lf(t)Lg(t)$$

(again this is an easy computation; if in doubt, check [33] for details). Let us now examine some special cases that we will need in Chapter 5. If a is a positive real number, let ϕ_a be the function defined by $\phi_a(t) := e^{-at}$. By a straightforward computation we have

$$(L\phi_a)(z) = \frac{1}{z + a}.$$

In the study of birth–death processes we will need the corresponding result for an arbitrary number of simple poles. Let us begin with the case of two factors: if a and b are two positive numbers with $a \neq b$, then you may check that

$$(\phi_a * \phi_b)(t) = \frac{e^{-bt} - e^{-at}}{a - b}.$$

This is therefore the inverse Laplace transform of the rational function $\frac{1}{(z+a)(z+b)}$; here is the n-term statement (see, for instance, [27], p. 224 or [33], p. 39):

A Some useful facts from calculus

Theorem A.10. *Let* $n \geq 2$ *be an integer, let* $a_1, \ldots, a_n$ *be* n *distinct positive numbers, and define*

$$F_n(z) := \prod_{k=1}^{n} \frac{1}{(z + a_k)};$$

then the inverse Laplace transform of F_n *is*

$$f_n(t) = \sum_{k=1}^{n} e^{-a_k t} \prod_{j=1, j \neq k}^{n} \frac{1}{(a_j - a_k)}. \qquad (A.9)$$

This means that the right-hand side of (A.9) is the convolution product $(\phi_{a_1} * \cdots * \phi_{a_n})(t)$.

Proof. The idea is to reduce the proof to the scalar case using partial fractions. The partial fraction decomposition of F_n is

$$F_n(z) = \frac{A_1}{z + a_1} + \cdots + \frac{A_n}{z + a_n},$$

where the (so-called) residue A_k is given by

$$A_k = \lim_{z \to -a_k} (z + a_k) F_n(z),$$

which means

$$A_k = \prod_{j=1, j \neq k}^{n} \frac{1}{a_j - a_k}.$$

The representation (A.9) follows immediately by linearity of the (inverse) Laplace transform. $\square$

The case in which all the a_k are equal will be needed in the study of the Poisson process; if G_n is defined by

$$G_n(z) := \frac{1}{(z + a)^n},$$

then it is easy to check that its inverse Laplace transform is given by

$$g_n(t) = \frac{t^{n-1}}{(n-1)!} e^{-at}. \qquad (A.10)$$

One way to show this is to check directly that the Laplace transform of g_n is G_n (in computing the integral you will stumble on the Euler gamma function; if necessary, see Appendix B for details); the other is to show that the n-fold convolution product $\phi_a * \cdots * \phi_a$ is equal to g_n, which is easily done by induction. This is an exercise in calculus; however, it has a probabilistic substance: it is equivalent to the statement that the sum of n exponential variables has a gamma distribution; again see Appendix B on the gamma distribution if more details are needed.

B

Some useful facts from probability

B.1 Some limit theorems in probability theory

B.1.1 Continuity of probability from above and below

We will repeatedly have to compute the probability of some event involving an infinite number of times by approximating it by events involving finitely many times only; the main technical device to do this is the so-called property of *continuity of probability*:

Theorem B.1. *1. (continuity from below of probability) If $A_1 \subset A_2 \subset \cdots \subset A_n \cdots$ is an increasing sequence of events, then $P(A_k)$ has a limit as $k \to \infty$, given by*

$$\lim_{k \to \infty} P(A_k) = P(\bigcup_{n=1}^{\infty} A_n).$$

2. (continuity from above of probability) If $A_1 \supset A_2 \supset \cdots \supset A_n \cdots$ is a decreasing sequence of events, then $P(A_k)$ has a limit as $k \to \infty$, given by

$$\lim_{k \to \infty} P(A_k) = P(\bigcap_{n=1}^{\infty} A_n).$$

Proof. To prove continuity from below, we write A_n and $\bigcup_{j=1}^{\infty} A_j$ as disjoint unions:

$$A_n = A_1 \cup (A_2 \setminus A_1) \cup \cdots \cup (A_n \setminus A_{n-1}),$$

$$\bigcup_{j=1}^{\infty} A_j = A_1 \cup (A_2 \setminus A_1) \cup \cdots \cup (A_j \setminus A_{j-1}) \cup \cdots.$$

By σ-additivity we obtain

© Springer International Publishing AG, part of Springer Nature 2018
J.-F. Collet, *Discrete Stochastic Processes and Applications*,
Universitext, https://doi.org/10.1007/978-3-319-74018-8

$$P(\bigcup_{j=1}^{\infty} A_j) = P(A_1) + \sum_{j=2}^{\infty} P(A_j \setminus A_{j-1})$$

$$= \lim_{n\to\infty} [P(A_1) + \sum_{j=2}^{n} P(A_j \setminus A_{j-1})] = \lim_{n\to\infty} P(A_n).$$

Continuity from above follows immediately if we apply continuity from below to the decreasing sequence $(\Omega \setminus A_n)_{n\in\mathbb{N}}$:

$$P(\Omega \setminus \bigcap_{n=1}^{\infty} A_n) = P(\bigcap_{n=1}^{\infty} [\Omega \setminus A_n]) = \lim_{n\to\infty} P(\Omega \setminus A_n).$$

□

A first consequence that we will need in Chapter 4 is the right continuity of the cumulative distribution function of a random variable:

Corollary B.2. *For a random variable X, its cumulative distribution function F defined by $F_X(t) := P(X \leq t)$ is a right continuous function. More generally, every function of t of the form $P(X \leq t, A|B)$, where A and B are two events with $P(B) \neq 0$, is right continuous.*

Proof. For a sequence (t_n) decreasing to t, the event $\{X \leq t\}$ may be represented as the decreasing intersection

$$\{X \leq t\} = \bigcap_{n\geq 1} \{X \leq t_n\}.$$

Therefore, continuity from above tells us that $F_X(t) = \lim_{n\to\infty} F_X(t_n)$, which exactly means that F_X is right continuous. The very same argument may be applied to the function $P(X \leq \cdot, A|B)$. □

Continuity from above is often used in the case that the complete intersection has zero probability. Here is a typical example, which we use in Chapter 4:

Corollary B.3. *Let X be a random variable satisfying $P(X > 0) = 1$. Then*
$$\lim_{h\to 0^+} P(X < h) = 0.$$

Proof. For any sequence h_k decreasing to zero, continuity from above applied to the decreasing sequence of events $\{X < h_k\}$ immediately gives the result. □

An equivalent statement near ∞ is the following:

Corollary B.4. *Let X be a random variable satisfying $P(X < \infty) = 1$. Then*

$$\lim_{n\to\infty} P(X \leq n) = 1.$$

Proof. Apply continuity from below to the sequence $\{X \leq n\}$, or apply the previous result to $\frac{1}{X}$. □

B.1.2 Three notions of convergence of sequences of random variables

If $(X_n)_{n \in \mathbb{N}}$ is a sequence of random variables on the same probability space, we say that

1. X_n converges to X *in probability* if for every $\epsilon > 0$,
$$\lim_{n \to \infty} P(|X_n - X| > \epsilon) = 0;$$

2. X_n converges to X *almost surely* if
$$P(\lim_{n \to \infty} X_n = X) = 1;$$

3. X_n converges to X *in distribution* if for every x,
$$\lim_{n \to \infty} P(X_n \leq x) = P(X \leq x).$$

The corresponding notation is respectively

$$X_n \overset{p}{\to} X, \quad X_n \overset{a.s.}{\to} X, \quad X_n \overset{d}{\to} X.$$

Convergence in distribution is just pointwise convergence of the cumulative distribution functions; note that if the probability space is discrete, this is equivalent to convergence of the probability mass function, i.e., to the requirement that
$$\forall x \in \mathcal{S} : P(X_n = x) \to P(X = x).$$

It is well known that almost sure convergence implies convergence in probability; conversely, the well-known monotone convergence theorem asserts that if a sequence of variables $(X_n)_{n \in \mathbb{N}}$ is increasing (meaning $X_n(\omega) \leq X_{n+1}(\omega)$ for all ω) and converges to X in probability, then it also converges to X almost surely.

B.2 Exponential and related distributions

B.2.1 Some properties of the exponential distribution

Recall that for $\lambda > 0$, the exponential distribution is defined over $[0, \infty)$ by the density $f_\lambda(t) = \lambda e^{-\lambda t}$; we write $T \sim \mathcal{E}(\lambda)$ to indicate that a variable T has this distribution; you should have no difficulty checking that in this case, the expected value of T is $\frac{1}{\lambda}$. In the study of the Poisson process we shall need the following two properties of independent exponentials:

Proposition B.5. *If $X \sim \mathcal{E}(\lambda)$ and $Y \sim \mathcal{E}(\mu)$ are two independent exponential variables, then*

B Some useful facts from probability

1. $\lambda X \sim \mathcal{E}(1)$;
2. $\min(X, Y) \sim \mathcal{E}(\lambda + \mu)$,
3. $P(X < Y) = \frac{\lambda}{\lambda + \mu}$.

The first two facts are easily established by determining the cumulative distribution functions (of λX and $\min(X, Y)$ respectively), and the third one by computing the integral of the joint density $f_{X,Y}(x, y) = \lambda e^{-\lambda x} \mu e^{-\mu y}$ over the set $\{y \in [0, \infty), 0 < x < y\}$. In investigating the explosion time of a continuous-time Markov chain in Chapter 4, we will require the following result on sums of exponential variables:

Lemma B.6. *Let $(X_k)_{k \in \mathbb{N}}$ be a collection of i.i.d. $\mathcal{E}(1)$ variables; then their sum almost surely converges to infinity:*

$$\sum_{k=1}^{n} X_k \overset{a.s.}{\to} \infty \quad \text{as} \quad n \to \infty.$$

Proof. Let $M > 0$ be fixed; for every $n \geq 2M$, writing $S_n := \sum_{k=1}^{n} X_k$, we have the inclusion of events

$$\{S_n \leq M\} \subset \{S_n - n \leq -\frac{n}{2}\} \subset \{|S_n - n| \geq \frac{n}{2}\};$$

hence

$$P(S_n \leq M) \leq P(|S_n - n| \geq \frac{n}{2}).$$

Using Chebyshev's inequality (note that $E(S_n) = n$ and $var(S_n) = n$), we obtain

$$P(S_n \leq M) \leq \frac{n}{\frac{n^2}{4}} = \frac{4}{n}.$$

This shows that S_n converges to ∞ in probability. Thus by the monotone convergence theorem, it also converges almost surely. □

B.2.2 The Gamma distribution

If $T_1, \ldots, T_k$ are k i.i.d. variables of exponential variables with parameter λ, then the sum $T_1 + \cdots + T_k$ has the following density:

$$f_{k,\lambda}(t) = \lambda^k \frac{t^{k-1}}{(k-1)!} e^{-\lambda t}, \quad t > 0. \tag{B.1}$$

This can be checked directly by convolution. More precisely, you should try to show by induction that

$$f_{k,\lambda} = f_\lambda * \cdots * f_\lambda,$$

with k terms on the right-hand side. Relation (B.1) defines the so-called $\Gamma(k, \lambda)$ distribution (sometimes also called the *Erlang distribution*), here presented in the shape-rate parametrization. The obvious generalization of (B.1) to the case in which k is not an integer is obtained by replacing the factorial with the Γ function:

$$f_{k,\lambda}(t) = \lambda^k \frac{t^{k-1}}{\Gamma(k)} e^{-\lambda t}, \quad t > 0.$$

This makes sense as soon as k is a positive number, and definition (A.6) immediately shows that this is indeed a probability density; however, when k is not an integer, the interpretation as a density of a sum of exponential variables is lost.

B.2.3 The truncated exponential distribution

In our study of the "bus paradox" in Chapter 3 we come across the truncated exponential distribution, which is defined and characterized as follows:

Definition B.7. *A* truncated exponential variable *is a variable of the form* $Y = \min(X, t)$, *where X is an exponential variable of parameter $\lambda > 0$, for some $\lambda > 0$, $t > 0$. In this case, we write $Y \sim \mathcal{TE}(\lambda, t)$.*

It is immediate to check that the cumulative distribution of Y is given by

$$P(Y \leq s) = \begin{cases} 1 & \text{for} \quad s \geq t, \\ 1 - e^{-\lambda s} & \text{for} \quad s < t. \end{cases}$$

Equivalently, this means that the probability measure of Y is given by

$$P_Y = \lambda e^{-\lambda s} \mathbb{1}_{[0,t]} dx + e^{-\lambda t} \delta_t$$

(where dx is the Lebesgue measure on $\mathbb{R}$, and δ_t is the Dirac measure at t). For statements that do not involve measure theory, this is also equivalent to the fact that for every interval A, the probability $P(Y \in A)$ is given by

$$P(Y \in A) = e^{-\lambda t} \mathbb{1}_A(t) + \int_{A \cap [0,t]} \lambda e^{-\lambda s} \, ds,$$

and also that for every test function α we have

$$E(\alpha(Y)) = e^{-\lambda t} \alpha(t) + \int_0^t \alpha(s) \lambda e^{-\lambda s} \, ds. \tag{B.2}$$

Note that Y is a variable that is neither discrete nor absolutely continuous: in the definition of Y the whole mass of the exponential variable X beyond t has been transferred to the single point t. This is accounted for by the first term in the above expression for $P(Y \in A)$ or $E(\alpha(Y))$.

B Some useful facts from probability

B.2.4 Binomial coefficients, binomial and related distributions

B.2.5 The Vandermonde convolution identity

In a few places in the text we make use of the following relation, known as *Vandermonde's identity*:

$$\binom{n+m}{l} = \sum_{k=0}^{l} \binom{n}{k}\binom{m}{l-k}. \tag{B.3}$$

Proof. Depending on your taste, this may be proved algebraically or combinatorially. For the algebraic proof we use the identity

$$(1+x)^{n+m} = (1+x)^n (1+x)^m.$$

Expanding all three terms by the binomial formula, we obtain

$$\sum_{l=0}^{n+m} \binom{n+m}{l} x^l = \left[\sum_{k=0}^{n} \binom{n}{k} x^k\right]\left[\sum_{j=0}^{m} \binom{m}{j} x^j\right] = \sum_{k=0}^{n} \sum_{l=k}^{n+m} \binom{n}{k}\binom{m}{l-k} x^l,$$

where the last expression was obtained by the change of summation index $l = k + j$. Using Fubini's theorem (draw a picture if necessary), this last sum may be rewritten as follows:

$$\sum_{k=0}^{n} \sum_{l=k}^{n+m} \binom{n}{k}\binom{m}{l-k} x^l = \sum_{l=0}^{n+m} \left[\sum_{k=0}^{l} \binom{n}{k}\binom{m}{l-k}\right] x^l.$$

Identification with the expansion of $(1+x)^{n+m}$ yields (B.3). To prove (B.3) combinatorially, note that the general term on the right-hand side counts the number of ways to draw k of objects sequentially from a pool of n objects and then $l - k$ objects from another pool of m objects; summing over k, all we did was select l objects from $l + m$ objects (the union of both pools). $\square$

B.2.6 Obtaining the Poisson distribution from the Binomial distribution

We all know that the binomial distribution $\mathcal{B}(n, p)$ has expectation np. If we want to let n get extremely large while maintaining this expectation finite, it is natural to take $p = \frac{n}{\lambda}$, where λ is a positive constant. In the language of Bernoulli trials, this means that we play an infinitely large number of times, with a vanishingly small chance of success. The limiting distribution in these asymptotics turns out to be the Poisson distribution of parameter λ:

Lemma B.8. *Let $\lambda > 0$ be fixed, and let X_n be a sequence of binomial variables with $X_n \sim \mathcal{B}(n, \frac{\lambda}{n})$. Then as $n \to \infty$, the sequence (X_n) converges in distribution to a Poisson variable of parameter λ:*

$$X_n \xrightarrow{d} X \sim \mathcal{P}(\lambda).$$

Proof. Since we are dealing with variables taking values in $\mathbb{N}$, it suffices to show that

$$\forall k \in \mathbb{N}: \quad \lim_{n \to \infty} P(X_n = k) = e^{-\lambda} \frac{\lambda^k}{k!}.$$

This is immediate using the fact that $(1 - \frac{\lambda}{n})^n \to e^{-\lambda}$.

$\square$

B.2.7 The negative binomial distribution

If we toss a coin that gives success with probability p (some people call this an infinite sequence of independent Bernoulli trials), our first success occurs at trial number n if and only if we obtained a sequence of $n-1$ failures followed by a success. If we denote by X the number of the trial that brings our first success, this means that for all $n \geq 1$ we have

$$P(X = n) = p(1 - p)^{n-1}.$$

This is the famous *geometric distribution of parameter p*. In this case, we shall write $X \sim \mathcal{G}(p)$. The fact that all probabilities $P(X = k)$ for $k \geq 1$ add up to 1 is a consequence of the formula for the geometric series,

$$\frac{1}{1 - t} = 1 + t + \cdots + t^n + \cdots, \tag{B.4}$$

applied to $t = 1 - p$. The *negative binomial distribution* appears when we consider further successes; for $k \geq 1$, the number Y of the trial that brings our kth success is distributed as follows:

$$P(Y = n) = \binom{n-1}{k-1} p^k (1 - p)^{n-k}, \quad n \geq k.$$

The abbreviated notation to indicate that Y follows the *negative binomial distribution of parameters k and p* is $Y \sim \mathcal{NB}(k, p)$. How do we make sure that all probabilities add up to 1? If we differentiate (B.4) k times, after some elementary manipulation we obtain

$$\frac{1}{(1 - t)^{k+1}} = \sum_{n=0}^{\infty} \binom{n}{k} t^{n-k}.$$

This (when applied to $t = 1 - p$) immediately yields

$$\sum_{n=k}^{\infty} P(Y = n) = 1.$$

B.3 Order statistics

If $X := (X_1, \ldots, X_n)$ is a random vector, for each $\omega \in \Omega$ we may order the collection of the values $X_1(\omega), \ldots, X_n(\omega)$ in a nondecreasing fashion. If we denote by $X_{(1)}(\omega), \ldots, X_{(n)}(\omega)$ the corresponding value, this gives us a random vector $(X_{(1)}, \ldots, X_{(n)})$, whose components are called the *order statistics* of X. In particular, the first and last order statistics are the minimum and maximum of the components of X:

$$X_{(1)} = \min_i X_i, \quad X_{(n)} = \max_i X_i.$$

In the study of the Poisson process we will require the following result on the distribution of the order statistics of a random sample (for a proof, see, for instance, [4], Proposition 13.15, p. 285):

Theorem B.9. *If the X_i are i.i.d. continuous variables with common density f, then the joint density of all the order statistics of X is given by*

$$f_{(X_{(1)}, \ldots, X_{(n)})}(x_1, \ldots, x_n) = n! f(x_1) \cdots f(x_n) \mathbb{1}_{x_1 < x_2 < \cdots < x_n}.$$

C

Some useful facts from linear algebra

Summary. In this appendix (assuming here that you have at least a working knowledge of the notion of eigenvalues and eigenvectors) I collect without proof the basic facts about linear algebra that are used in Chapter 2; for more on these topics, see, for instance, [23], [15], or [25]. Denote by $\mathcal{M}_n(\mathbb{R})$ the set of all $n \times n$ square matrices with real entries. The main raison d'être for matrices is obviously to multiply vectors; due to the fact that real matrices may have nonreal eigenvalues, it is sometimes necessary to consider the action of a matrix $A \in \mathcal{M}_n(\mathbb{R})$ on $\mathbb{C}^n$ rather than just $\mathbb{R}^n$. Therefore, at the risk of sounding slightly pedantic, we have to consider $\mathcal{M}_n(K)$, where K may be either $\mathbb{R}$ or $\mathbb{C}$.

C.1 Matrix norms

Definition C.1. *A matrix norm on $\mathcal{M}_n(K)$ is a map $|\cdot|$ that to every matrix $A \in \mathcal{M}_n(K)$ assigns a nonnegative number $|A|$ and enjoys the following properties:*

separation: $\quad |A| = 0$ *only if* $A = 0$;
homogeneity: $\quad \forall \lambda \in K, A \in \mathcal{M}_n(K) : |\lambda A| = |\lambda| A$;
subadditivity: $\quad \forall A, B \in \mathcal{M}_n(K) : |A + B| \leq |A| + |B|$;
submultiplicativity: $\quad \forall A, B \in \mathcal{M}_n(K) : |AB| \leq |A||B|$.

The first three requirements tell us that the map $|\cdot|$ is a norm on the vector space $\mathcal{M}_n(K)$ (see, for instance, [25] if a brushup on vector norms is needed).

The *Frobenius norm* (also called the *Hilbert–Schmidt norm* or *Schur norm*) of a matrix is defined by

$$|A|_F := \left(\sum_{i=1}^{n}\sum_{j=1}^{n}|a_{ij}|^2\right)^{\frac{1}{2}}. \tag{C.1}$$

It is easy to check that this is indeed a matrix norm on $\mathcal{M}_n(K)$.

The *conjugate matrix* A^* of a matrix $A \in \mathcal{M}_n(\mathbb{C})$ is the complex conjugate transpose of A:

© Springer International Publishing AG, part of Springer Nature 2018
J.-F. Collet, *Discrete Stochastic Processes and Applications*,
Universitext, https://doi.org/10.1007/978-3-319-74018-8

C Some useful facts from linear algebra

$$A^* = \bar{A}^t.$$

A matrix $A \in \mathcal{M}_n(\mathbb{C})$ is said to be *unitary* if it is invertible and its inverse is its conjugate:

$$AA^* = A^*A = I_n.$$

Unfortunately (like many other areas of mathematics, elementary linear algebra suffers from an overabundance of equivalent terms), if a matrix A has real entries, then the word *orthogonal* is preferred: $A \in \mathcal{M}_n(\mathbb{R})$ is said to be orthogonal if

$$AA^t = A^tA = I_n.$$

The Frobenius norm is invariant under multiplication by unitary matrices: if A is any complex matrix and U is a unitary matrix, then

$$|AU|_F = |A|_F.$$

A given norm $|\cdot|$ on K^n may be used to define a norm on $\mathcal{M}_n(K)$ as follows (note that here we are using the same notation $|\cdot|$ whether the argument is a vector or a matrix): if for $A \in \mathcal{M}_n(K)$ we put

$$|A| := \sup_{x \in K^n} \frac{|Ax|}{|x|},$$

then it is easy to check that this does indeed define a matrix norm on $\mathcal{M}_n(K)$, which is said to be *induced by*, or *subordinate to*, the norm on K^n.

The two most important induced norms on $\mathcal{M}_n(K)$ are the following:

Theorem C.2. *On $\mathbb{C}^n$, the ∞-norm and 1-norm are defined by*

$$\forall x \in \mathbb{C}^n : \quad |x|_\infty := \max_i |x_i|, \quad |x|_1 := \sum_i |x_i|.$$

The induced norms on $\mathcal{M}_n(K)$ are respectively the so-called max-absolute-row-sum norm,

$$|A|_\infty = \max_i \sum_j |a_{ij}|,$$

and max-absolute-column-sum norm,

$$|A|_1 = \max_j \sum_i |a_{ij}|.$$

C.2 Eigenvalues and spectral radius

The Schur triangularization theorem (see, for instance, [25], p. 508 for a proof) tells us that every square matrix is unitarily similar to an upper triangular matrix:

Theorem C.3 (Schur's triangularization theorem). *For every square matrix $A \in \mathcal{M}_n(\mathbb{C})$ there exists a unitary matrix $U \in \mathcal{M}_n(\mathbb{C})$ such that the matrix U^*AU is upper triangular.*

Note that in this case, the matrices A and U^*AU must have the same characteristic polynomial. This implies that the diagonal of the matrix U^*AU is made up of the eigenvalues of A, and for each $\lambda \in \sigma(A)$ the number of times it appears on the diagonal is its algebraic multiplicity. An important consequence of Schur's theorem is that it tells us what the spectrum becomes when we compute powers, or more generally polynomials, of a square matrix:

Theorem C.4. *For every square matrix $A \in \mathcal{M}_n(\mathbb{C})$ and complex-valued polynomial P, the spectrum of $P(A)$ is simply the set*

$$\sigma(P(A)) = \{P(\lambda), \; \lambda \in \sigma(A)\}.$$

Indeed, it suffices to notice that if $U^*AU = T$ is the Schur form of A, then $U^*P(A)U = P(T)$ is also upper triangular.

How is the spectral radius related to matrix norms? First, for every subordinate norm we have the bound

$$\rho(A) \leq |A| \tag{C.2}$$

(to prove this, choose an eigenvector x and bound the quantity $|Ax|$). Second, the *spectral radius formula*, a.k.a. Gelfand's formula, asserts that for every matrix norm, the spectral radius $\rho(A)$ is given by

$$\rho(A) = \lim_{k \to \infty} |A^k|^{\frac{1}{k}}. \tag{C.3}$$

Finally, we will need the notion of *convergent matrix*, as well as its characterization in terms of the spectral radius:

Theorem C.5. *For a square matrix A, the limit of A^k as $k \to \infty$ is 0 if and only if $\rho(A) < 1$; such matrices are called convergent matrices.*

For a proof, see, for instance, [25], p. 617. Note that we do not need to be any more precise when we say that A^k converges to 0: since the set of $n \times n$ matrices is finite-dimensional, all norms are equivalent, and this convergence is equivalent to the convergence of $A^k(i,j)$ to 0 for all i,j. Also note that some texts use a slightly different terminology and use *convergent* to describe any matrix A such that A^k has a limit. Here is why convergent matrices are so important:

Theorem C.6. *If A is a convergent matrix, then the matrix $I - A$ is invertible, and its inverse is given by the sum of the series*

$$(I - A)^{-1} = I + A + \cdots + A^n + \cdots.$$

Proof. Let $\epsilon > 0$ be small enough so as to have $\rho(A) + \epsilon < 1$; the spectral radius formula tells us that for k large enough, we have $|A^k| \leq (\rho(A) + \epsilon)^k$, which means that our series is normally convergent. Now, for every n, the partial sum of the series commutes with $I - A$, and

$$(I - A)(I + A + \cdots + A^n) = I - A^{n+1}.$$

The right-hand side converges to I as $n \to \infty$, which shows that the sum S of the series satisfies $S(I - A) = (I - A)S = I$. $\square$

In Chapter 2 we make use of the power method, which is based on the following two results:

Theorem C.7. *Let λ be a simple eigenvalue of a square matrix A, and let x, y be respectively right and left eigenvectors associated with λ. Then $yx \neq 0$, and the projector onto $N(A - \lambda I)$ along $R(A - \lambda I)$ is $\frac{xy}{yx}$.*

(For a proof, see [25], p. 518.)

Theorem C.8. *If $\rho(A) \neq 0$ is a dominant eigenvalue of A, then as $k \to \infty$, the matrix $(\frac{A}{\rho(A)})^k$ converges to the projector onto $N(A - \rho(A)I)$ along $R(A - \rho(A)I)$.*

Proof. This is a special case of a more general result that you will find in [25] (see 7.10.34 on p. 630). Let us give a simple proof based on Theorem C.5. Let Π denote the projector as in the statement of the result; recall that Π commutes with A. Writing any x as $x = \Pi x + (I - \Pi)x$ and applying $(\frac{A}{\rho(A)})^k$, we obtain

$$\left(\frac{A}{\rho(A)}\right)^k = \Pi + [\frac{A}{\rho(A)}(I - \Pi)]^k. \tag{C.4}$$

If $\mu \neq 0$ is an eigenvalue of $A(I - \Pi)$ and x is an associated eigenvector, applying Π to the relation $A(I - \Pi)x = \mu x$, we obtain $x = (I - \Pi)x$, and thus $Ax = \mu x$. Since $\rho(A)$ is dominant, this implies $\mu < \rho(A)$. This means that the spectral radius of $\frac{A}{\rho(A)}(I - \Pi)$ is strictly less than 1, so by Theorem C.5, the second term in (C.4) converges to 0. $\square$

C.3 Monotone matrices and M matrices

Results of Perron and Frobenius (Theorems 2.15 and 2.27 in Chapter 2) highlight the importance of nonnegative matrices. In this connection it seems well worth investigating the class of invertible matrices with a nonnegative inverse. Here is a nice characterization, Collatz's theorem:

Theorem C.9. *For a square matrix A with real entries, the following two conditions are equivalent:*

$$A \text{ is invertible and } A^{-1} \geq 0, \tag{C.5}$$

$$Ax \geq 0 \implies x \geq 0. \tag{C.6}$$

Proof. The second condition is immediate from the first, as can be seen by multiplying the inequality $Ax \geq 0$ by A^{-1}. Conversely, if the second condition is true, then $Ax = 0$ implies that both Ax and $A(-x)$ are nonnegative, and thus using the condition twice, we get $x \geq 0$ and $-x \geq 0$, thus $x = 0$, which means that A is invertible. To show that $A^{-1} \geq 0$, denoting the ith basis vector by e_i, we note that $AA^{-1}e_i = e_i \geq 0$ implies that the ith column vector $A^{-1}e_i$ of the matrix A^{-1} is nonnegative, i.e., $A^{-1} \geq 0$. □

This motivates the following definition:

Definition C.10. *A square matrix A is said to be* monotone *if*

$$Ax \geq 0 \implies x \geq 0.$$

How can we prove that a given square matrix is monotone? As in Theorem 2.2 and Lemma 2.8, the "weight" of the diagonal of A plays an important role. Here is the relevant notion:

Definition C.11. *A square matrix A is said to be* diagonally dominant *if*

$$\forall i : \quad |A(i,i)| \geq \sum_{j \neq i} |A(i,j)|;$$

it is said to be strictly diagonally dominant *if all inequalities are strict.*

If we add a condition on the signs of the entries, this turns out to ensure monotonicity:

Theorem C.12. *Assume that A is strictly diagonally dominant, that all its diagonal terms are positive, and that all its off-diagonal terms are nonpositive:*

$$\forall i : \quad A(i,i) > 0, \quad \forall j \neq i : \quad A(i,j) \leq 0.$$

Then A is monotone.

Proof. Let D be the diagonal matrix made up of the diagonal elements of A, i.e., $D(i,i) = A(i,i)$. Obviously, D is monotone, and writing

$$A = D - (D - A) = D[I - D^{-1}(D - A)],$$

it suffices to show that $I - D^{-1}(D - A)$ is monotone. Putting $B := D^{-1}(D-A)$, in view of Theorem C.9 we need to show that $I - B$ is invertible and has a nonnegative inverse. Note that $B > 0$, as a consequence of the fact that A is strictly diagonally dominant. Therefore, in view of Theorem C.6, it suffices to show that the series $I + B + \cdots + B^n + \cdots$ converges; this will imply both existence and positivity of the inverse $(I - B)^{-1}$. We shall now show that $\rho(B) < 1$ by considering the row sums of B. Dividing the inequality

$$\sum_{j \neq i} |A(i,j)| < A(i,i)$$

by $A(i,i)$, we obtain

$$\sum_{j\neq i} |B(i,j)| < 1,$$

and thus by Theorem 2.13, we have $\rho(B) < 1$. □

Closely related to this result is the notion of M matrices:

Definition C.13. *A square matrix A is said to be an M matrix if it is monotone and all its off-diagonal elements are nonpositive: $A(i,j) \leq 0$ for $i \neq j$.*

It is quite easy to find monotone matrices that are not M matrices (see, for instance, Problem 2.5 in Chapter 2 for a characterization of 2×2 monotone matrices that are M matrices). For various characterizations of M matrices, see [25], pp. 626 and 639. The following is an important consequence of Theorem C.12 that we will need in Chapter 2 for the study of the PageRank algorithm:

Corollary C.14. 1. *Let B be a nonnegative matrix. Then for all $r > \rho(B)$, the matrix $A := rI - B$ is an M matrix.*
2. *As a consequence, if P is a Markov matrix, then for all $\alpha \in (0,1)$, the matrix $I - \alpha P$ is an M matrix.*

Proof. The off-diagonal elements of A are clearly negative; thus we need to show only that A is monotone, which by Theorem C.9 means showing that it is invertible with nonnegative inverse. Dividing by r, let us consider instead the matrix $I - \frac{B}{r}$. Since $r > \rho(B)$, we have $\rho(\frac{B}{r}) < 1$, and therefore, as in the proof of Theorem C.12, this implies that the matrix $I - \frac{B}{r}$ is invertible, and by Theorem C.6, its inverse is given by the sum of a matrix series in which all terms are nonnegative. This means that $I - \frac{B}{r}$ is monotone, and therefore so is A. The second point follows immediately, since by Theorem 2.1, we know that $\rho(P) = 1$. □

For the sake of completeness let us mention that the converse of the first point in the conclusion is also true, that is, that every M matrix A has a representation of the form $A = rI - B$ with $B \geq 0$ and $r > \rho(B)$ (see, for instance, [25], p. 639, for a proof). Indeed, in some texts it is this representation that is taken as a definition of M matrices.

C.4 Permutation matrices

In Chapter 2 we investigate the properties of the directed graph $G(A)$ associated with a square matrix A; the relevant properties of the graph (such as strong connectivity or lack thereof) are not exactly intrinsic to the matrix, since the definition of $G(A)$ uses a numbering of the vertices. To put it bluntly, it does not matter how you label two vertices; what matters is whether they are connected. The mathematical concept relevant here is that of *permutation matrix.* If $\sigma \in \mathfrak{S}_n$ is a permutation, we can use σ to define a linear map L_σ on vectors by simply permuting the entries. More precisely, for $x \in \mathbb{R}^n$, the vector $L_\sigma(x)$ is defined by $(L_\sigma(x))(j) = x(\sigma(j))$. The permutation matrix P_σ is the associated matrix acting on row vectors. This motivates the following definition:

Definition C.15. *For a permutation $\sigma \in \mathfrak{S}_n$, the permutation matrix P_σ is the matrix defined by*

$$P_\sigma(i,j) := \delta_{i\sigma(j)}.$$

You may check that this is consistent with our definition of the map L_σ. Note that in some texts, the map L_σ is not used explicitly, and a permutation matrix is just defined as a matrix that in each row and each column has exactly one nonzero entry, which is 1.

We leave it to you as an (easy) exercise to check the following.

Theorem C.16. *Some algebraic properties of permutation matrices are as follows:*

1. product:

$$P_{\sigma_1} P_{\sigma_2} = P_{\sigma_2} P_{\sigma_1} = P_{\sigma_1 \sigma_2};$$

2. inverse:

$$P_\sigma^{-1} = P_\sigma^t = P_{\sigma^{-1}};$$

3. multiplication of square matrices:

$$(P_\sigma A)(i,j) = A(\sigma^{-1}(i), j); \quad (AP_\sigma)(i,j) = A(i, \sigma(j));$$

4. permutation similarity:

$$(P_\sigma^t A P_\sigma)(i,j) = A(\sigma(i), \sigma(j)). \tag{C.7}$$

At this point you should try to play with a simple permutation, say of three elements, and check by an example the formulas in item 3 above (the first one, for example, says that row k of the matrix A becomes row $\sigma(k)$ of the matrix $P_\sigma A$; what is the corresponding statement for columns?).

C.5 Matrix exponentials

The most elementary (and classical) way to define the exponential of a square matrix is based on the power series expansion of the exponential function for complex numbers:

$$\forall z \in \mathbb{C}: \quad e^z = 1 + z + \frac{z^2}{2} + \cdots + \frac{z^n}{n!} + \cdots .$$

Here is the precise statement:

C Some useful facts from linear algebra

Theorem C.17. *For an $n \times n$ square matrix A with complex entries, the series*

$$I_n + A + \frac{A^2}{2} + \cdots + \frac{A^k}{k!} + \cdots$$

is normally convergent in every matrix norm. Generalizing the notation used for $n = 1$, we denote by e^A the sum of this series. Then:

1. *If B is another $n \times n$ matrix such that $AB = BA$, we have $e^{A+B} = e^A e^B = e^B e^A$.*
2. *As a consequence, e^A is invertible, and its inverse is e^{-A}.*

Note that since e^A is defined as a limit of polynomials in A, it commutes with A; more generally, if A and B commute, then so do e^A and every polynomial in B.

Proof. The fact that the series is normally convergent comes from the bound $|A^k| \leq |A|^k$, so our matrix exponential is well defined as its sum. To prove its algebraic properties, note that since A and B commute, we may apply the binomial formula to expand all powers of $A + B$; therefore,

$$e^{A+B} = \sum_{n=0}^{\infty} \frac{(A+B)^n}{n!} = \sum_{n=0}^{\infty} \frac{1}{n!} \sum_{k=0}^{n} \binom{n}{k} A^k B^{n-k}$$

$$= \sum_{k=0}^{\infty} \sum_{n=k}^{\infty} \frac{1}{k!(n-k)!} A^k B^{n-k} = \sum_{k=0}^{\infty} \frac{1}{k!} A^k \sum_{l=0}^{\infty} \frac{1}{l!} B^l$$

$$= e^A e^B.$$

$\square$

One of the main uses of matrix exponentials is the resolution of linear differential equations (in $\mathbb{R}^n$) with constant coefficients. Let us begin with a basic differentiability result:

Lemma C.18. *For every $n \times n$ matrix A, the map $t \in \mathbb{R} \mapsto e^{tA}$ is differentiable everywhere on $\mathbb{R}$, and*

$$\frac{d}{dt} e^{tA} = A e^{tA} = e^{tA} A.$$

Proof. Since the radius of convergence of the power series is infinite, we may differentiate it termwise, which immediately gives the result. $\square$

An interesting consequence of commuting matrices is the following:

Lemma C.19. *For two $n \times n$ matrices A and B, the following three conditions are equivalent:*

$$AB = BA; \tag{C.8}$$

$$\forall t \in \mathbb{R}: \quad B e^{tA} = e^{tA} B; \tag{C.9}$$

$$\exists \epsilon > 0 \quad \forall t \in (0, \epsilon): \quad B e^{tA} = e^{tA} B. \tag{C.10}$$

Proof. If A and B commute, then every polynomial in A commutes with B, which shows that e^{tA} and B commute for all $t \in \mathbb{R}$, so clearly (C.8) implies (C.9) and (C.10). To show that (C.10) implies (C.8), taking derivatives, we obtain for all $t \in (0, \epsilon)$ the relation

$$BAe^{tA} = Ae^{tA}B,$$

which on taking the limit $t \to 0$ yields $AB = BA$. □

Here comes an interesting application to linear ordinary differential equations with constant coefficients:

Theorem C.20. *For a given $n \times n$ matrix A, consider the following Cauchy problem in $\mathcal{M}_{p \times n}(\mathbb{C})$:*

$$\begin{cases} \frac{dX}{dt} = AX, \\ X(0) = X_0, \end{cases} \tag{C.11}$$

in which $X_0 \in \mathcal{M}_{p \times n}(\mathbb{C})$ is given. The unique solution to this Cauchy problem is given by

$$X(t) = e^{tA} X_0.$$

Similarly, if we take $Y_0 \in \mathcal{M}_{n \times p}(\mathbb{C})$, the Cauchy problem

$$\begin{cases} \frac{dY}{dt} = YA, \\ Y(0) = Y_0, \end{cases} \tag{C.12}$$

is well posed in $\mathcal{M}_{n \times p}(\mathbb{C})$ and has the unique solution $Y(t) = Y_0 e^{tA}$.

Proof. It follows immediately from Lemma C.18 that $e^{tA} X_0$ is a solution to (C.11). The right-hand side AX is a Lipschitz continuous function of X (whatever matrix norm we use), so by the well-known existence and uniqueness result on ordinary differential equations, our system (C.11) is well posed (see, for instance, [14] for details), which means that $e^{tA} X_0$ is the unique solution. However, in this simple case, uniqueness may be proved directly without having to appeal to the general theory. Indeed, if $X(t)$ is a solution, from (C.18) you can check that $\frac{d}{dt}[e^{-tA} X(t)]$ vanishes identically. This means that $e^{-tA} X(t)$ is constant, and thus using the initial condition, we see that $X(t) = e^{tA} X_0$. This proves the first half of our theorem regarding (C.11); the other half (regarding (C.12)) follows immediately by transposition. □

In the form (C.11), the two most common cases of application of this result are $p = 1$, in which we solve a linear system of n ordinary differential equations in n unknown functions, and $p = n$, in which case the solution to the problem is an $n \times n$ matrix. In this case we may ask whether the solution at an arbitrary time will commute with A if and only if it does so at time 0:

C Some useful facts from linear algebra

Corollary C.21. *Let A and X_0 be two given $n \times n$ matrices. Then the Cauchy problems*

$$\begin{cases} \frac{dX}{dt} = AX, \\ X(0) = X_0, \end{cases} \tag{C.13}$$

and

$$\begin{cases} \frac{dX}{dt} = XA, \\ X(0) = X_0, \end{cases} \tag{C.14}$$

are both well posed; they are equivalent (meaning both have the same solution) if and only if the matrices A and X_0 commute.

Proof. We already know that both problems are well posed, with respective solutions $e^{tA}X_0$ and $X_0 e^{tA}$, and we know from Lemma C.19 that these are equal for all t if and only if $AX_0 = X_0 A$. $\square$

In the framework of complex-valued functions of one real variable, it is well known that under suitable regularity assumptions, the exponential map e^z is the unique solution to the Cauchy problem $y' = y$, $y(0) = 1$. Let us now indicate a generalization of this result to matrix-valued maps:

Theorem C.22. *For some $T \leq \infty$, let $P : [0, T) \to \mathcal{M}_{n \times n}(\mathbb{C})$ be a matrix-valued map satisfying $P(0) = I_n$ and the so-called semigroup property:*

$$\forall t \geq 0, s > 0 : \quad P(t+s) = P(t)P(s). \tag{C.15}$$

Let us assume that P is right differentiable at 0, i.e., that the right derivative

$$A := \lim_{h \to 0^+} \frac{P(t) - I}{h}$$

exists; by this we mean that this limit exists componentwise, and that $A(i, j)$ is finite for all i, j. Then for all $t \geq 0$ we have $P(t) = e^{tA}$.

Proof. We are going to show that the map P is differentiable on the open interval $(0, T)$, and that, for all $t > 0$, we have

$$P'(t) = P(t)A = AP(t).$$

In view of Theorem C.20, this will yield the desired representation. First note that from (C.15) we have the commutation relation $P(s)P(t) = P(t)P(s)$. Since P is right differentiable at 0, it is right continuous at 0, i.e., $P(s) \to I$ as $s \to 0$. Therefore, $P(s)$ is invertible for $s > 0$ small enough; we may then use (C.15) to conclude that $P(t)$ is invertible for all $t \in [0, T)$.

Let us first prove continuity on the open interval. From (C.15) we see that for all $t > 0$, the map P is right continuous at t. To check left continuity, for $s > 0$ small enough we have $P(t-s)P(s) = P(t)$; hence $P(t-s) = P(t)P(s)^{-1}$

converges to $P(t)$ as $s \to 0^+$, which shows that P is left continuous, therefore continuous, at t. We may now show differentiability proceeding in a similar fashion, i.e., treating right and left differentiability separately. If $h > 0$, we have

$$\frac{P(t+h) - P(t)}{h} = P(t)\frac{P(h) - I}{h}.$$

As $h \to 0^+$ this converges to $P(t)A$, which shows that the right derivative at t exists and is equal to $P(t)A$. On the other hand, for $h < 0$ we may write

$$\frac{P(t+h) - P(t)}{h} = P(t+h)\frac{P(-h) - I}{-h}.$$

Now, as $h \to 0^-$ this converges to $P(t)A$, which shows that the left derivative at t exists and is equal to $P(t)A$. □

D

An arithmetic lemma

In the course of the proof of Theorem 1.48 we made use of the following fact:

Lemma D.1. *Let A be a set of relatively prime positive integers. Assume that A is closed under addition:*

$$x, y \in A \implies x + y \in A.$$

Then the complement of A in $\mathbb{N}$ is finite.

Let us see how this is proved. We will need the following:

Lemma D.2. *Let A be an infinite subset of $\mathbb{N}$ consisting of relatively prime integers; then there exists a finite subset of A with the same property.*

Proof. Pick some $x \in A$, and let $p_1, \ldots, p_k$ be the prime divisors of x; for every i between 1 and k, p_i cannot divide all elements of A, so let us choose some $x_i \in A$ that p_i does not divide. Then the numbers $x, x_1, \ldots, x_k$ are relatively prime (as an interesting exercise you may want to check that this fact generalizes to any greatest common divisor: every infinite set of positive integers has a finite subset with the same greatest common divisor.) $\square$

We can now prove Lemma D.1:
We begin by remarking that A is closed under multiplication by positive integers, and therefore closed under linear combinations with such coefficients:

$$x, y \in A, n_1, n_2 \in \mathbb{N} \implies n_1 x + n_2 y \in A.$$

Therefore, it suffices to show that every integer larger than some n_0 may be written as a linear combination of this type. From the previous lemma we may choose a finite collection $a_1, \ldots, a_r$ of relatively prime elements of A, and using Bézout's lemma, we have

$$t_1 a_1 + \cdots + t_r a_r = 1$$

D An arithmetic lemma

for some integers $t_i \in \mathbb{Z}$. In order to obtain an integer n as a linear combination of the a_i's we perform the Euclidian division of n by their sum:

$$n = k \sum_{i=1}^{r} a_i + s, \quad 0 \le s < \sum_{i=1}^{r} a_i.$$

Thus we obtain

$$n = k \sum_{i=1}^{r} a_i + s \sum_{i=1}^{r} a_i t_i = \sum_{i=1}^{r} (k + s t_i) a_i.$$

If n is large enough, all coefficients $k + s t_i$ in the above sum are positive, and this provides the desired representation for n. $\quad\square$

E

Table of exponential families

Table E.1. Exponential families: discrete distributions, part 1

Distribution	Binomial	Negative binomial	Geometric	Poisson
$P(X = x)$	$\binom{n}{x}p^x(1-p)^{n-x}$	$\binom{x+r-1}{x}p^x(1-p)^r$	$p(1-p)^{x-1}$	$e^{-\lambda}\frac{\lambda^x}{x!}$
$x \in$	$\{1, \cdots n\}$	$\mathbb{N}$	$\mathbb{N}^*$	$\mathbb{N}$
parameter	p	r	p	λ
$t(x)$	x	x	x	x
θ	$\log\frac{p}{1-p}$	$\log p$	$\log(1-p)$	$\log\lambda$
$F(\theta)$	$n\log(1+e^\theta)$	$-r\log(1-e^\theta)$	$-\log(e^{-\theta}-1)$	e^θ
$k(x)$	$\log\binom{n}{x}$	$\log\binom{x+r-1}{x}$	0	$-\log(x!)$
$E(X)$	np	$r\frac{p}{1-p}$	$\frac{1}{p}$	λ
$var(X)$	$np(1-p)$	$r\frac{p}{(1-p)^2}$	$\frac{1-p}{p^2}$	λ

© Springer International Publishing AG, part of Springer Nature 2018
J.-F. Collet, *Discrete Stochastic Processes and Applications,*
Universitext, https://doi.org/10.1007/978-3-319-74018-8

E Table of exponential families

Table E.2. Exponential families: discrete distributions, part 2

Distribution	Zipf	Zeta	Power series	Logarithmic
$P(X = x)$	$\frac{1}{Cx^s}, C := \sum_y y^{-s}$	$\frac{1}{\zeta(s)x^s}$	$u(x)\frac{1}{C\lambda^x}, C := \sum_y u(y)\lambda^y$	$-\frac{1}{\log(1-p)}\frac{p^x}{x}$
$x \in$	$\{1, \cdots N\}$	$\mathbb{N}^*$	$\mathbb{N}$	$\mathbb{N}^*$
parameter	s	s	λ	p
$t(x)$	$-\log x$	$-\log x$	x	x
θ	s	s	$\log \lambda$	$\log p$
$F(\theta)$	$\log\left(\sum_{x=1}^{N} x^{-s}\right)$	$\log \zeta(\theta)$	$\log \sum_y u(y)\lambda^y$	$\log(-\log(1-p))$
$k(x)$	0	0	$\log u(x)$	$-\log x$
$E(X)$	*	*	*	$-\frac{p}{1-p}\frac{1}{\log(1-p)}$
$var(X)$	*	*	*	$-p\frac{\log(1-p)-p}{(1-p)^2\log^2(1-p)}$

*Quantities for which no closed formula is available are indicated by a star.

Table E.3. exponential families: absolutely continuous distributions, part 1

Distribution	Exponential	Laplace	Weibull	Pareto
$f_X(x)$	$\lambda e^{-\lambda x}$	$\frac{\lambda}{2}e^{-\lambda|x|}$	$\alpha\lambda x^{\alpha-1}\exp(-\lambda x^\alpha)$	$k\frac{x_m^k}{x^{k+1}}$
$x \in$	$[0, \infty)$	$\mathbb{R}$	$[0, \infty)$	$[x_m, \infty)$
parameter	λ	λ	λ	k
$t(x)$	$-x$	$-|x|$	$-x^\alpha$	$-\log x$
θ	λ	λ	λ	k
$F(\theta)$	$-\log \theta$	$-\log \frac{\theta}{2}$	$-\log \theta$	$-\log(\theta x_m^\theta)$
$k(x)$	0	0	$(\alpha-1)\log x + \log \alpha$	$-\log x$
$E(X)$	$\frac{1}{\lambda}$	0	$\lambda^{-\frac{1}{k}}\Gamma(1+\frac{1}{\alpha})$	$\frac{k}{k-1}x_m, k > 1$
$var(X)$	$\frac{1}{\lambda^2}$	$\frac{2}{\lambda^2}$	$\lambda^{-\frac{2}{k}}[\Gamma(1+\frac{2}{\alpha}) - (\Gamma(1+\frac{1}{\alpha}))^2]$	$\frac{k}{(k-2)(k-1)^2}x_m^2, k > 2$

Table E.4. exponential families: absolutely continuous distributions, part 2

Distribution	Gamma	Inverse Gamma	Levy	Chi-sqare
$f_X(x)$	$\frac{\beta^\alpha}{\Gamma(\alpha)}x^{\alpha-1}\exp(-\beta x)$	$\frac{\beta^\alpha}{\Gamma(\alpha)}x^{-\alpha-1}\exp(-\frac{\beta}{x})$	$\sqrt{\frac{b}{2\pi}}\frac{1}{(x-\mu)^{\frac{3}{2}}}\exp(-\frac{b}{2(x-\mu)})$	$\frac{1}{2^{\frac{k}{2}}\Gamma(\frac{k}{2})}x^{\frac{k}{2}-1}\exp-(\frac{x}{2})$
$x \in$	$[0,\infty)$	$[0,\infty)$	$[\mu,\infty)$	$[0,\infty)$
parameter	β	β	b	k
$t(x)$	$-x$	$-\frac{1}{x}$	$-\frac{1}{2(x-\mu)}$	$\log x$
θ	β	β	b	$\frac{k}{2}$
$F(\theta)$	$-\alpha\log\theta$	$-\alpha\log\theta$	$-\frac{1}{2}\log\theta$	$\theta\log 2 + \log\Gamma(\theta)$
$k(x)$	$(\alpha-1)\log x - \log\Gamma(\alpha)$	$-(\alpha+1)\log x - \log\Gamma(\alpha)$	$-\frac{3}{2}\log(x-\mu) - \log\sqrt{2\pi}$	$-\log x - \frac{x}{2}$
$E(X)$	$\frac{\alpha}{\beta}$	$\frac{\beta}{\alpha-1}, \alpha > 1$	∞	k
$var(X)$	$\frac{\alpha}{\beta^2}$	$\frac{\beta^2}{(\alpha-2)(\alpha-1)^2}, \alpha > 2$	∞	$2k$

Table E.5. exponential families: absolutely continuous distributions, part 3

Distribution	Chi	Beta	Beta-prime	Gaussian
$f_X(x)$	$\frac{1}{2^{\frac{k}{2}-1}\Gamma(\frac{k}{2})}x^{k-1}\exp-(\frac{x^2}{2})$	$\frac{1}{B(\alpha,\beta)}x^{\alpha-1}(1-x)^{\beta-1}$	$\frac{1}{B(\alpha,\beta)}\frac{x^{\alpha-1}}{(1+x)^{\alpha+\beta}}$	$\frac{1}{\sqrt{2\pi\sigma^2}}\exp(-\frac{|x-\mu|^2}{2\sigma^2})$
$x \in$	$[0,\infty)$	$[0,1]$	$[0,\infty)$	$\mathbb{R}$
parameter	k	(α,β)	(α,β)	(μ,σ)
$t(x)$	$\log x$	$(\log x, \log(1-x))$	$(\log x, \log(1+x))$	$(-x^2,x)$
θ	k	(α,β)	$(\alpha,\alpha+\beta)$	$(\frac{1}{2\sigma^2},\frac{\mu}{\sigma^2})$
$F(\theta)$	$\frac{\theta}{2}\log 2 + \log\Gamma(\frac{\theta}{2})$	$\log B(\theta_1,\theta_2)$	$\log B(\theta_1,\theta_1+\theta_2)$	$\frac{\theta_2^2}{4\theta_1} - \frac{1}{2}\log\theta_1$
$k(x)$	$-\log(2x) - \frac{x^2}{2}$	$-\log(x(1-x))$	$-\log x$	$-\log\sqrt{2\pi}$
$E(X)$	$\sqrt{2}\frac{\Gamma(\frac{k+1}{2})}{\Gamma(\frac{k}{2})}$	$\frac{\alpha}{\alpha+\beta}$	$\frac{\alpha}{\beta-1}, \beta > 1$	μ
$var(X)$	$k - 2[\frac{\Gamma(\frac{k+1}{2})}{\Gamma(\frac{k}{2})}]^2$	$\frac{\alpha\beta}{(\alpha+\beta+1)(\alpha+\beta)^2}$	$\frac{\alpha(\alpha+\beta-1)}{(\beta-2)(\beta-1)^2}, \beta > 2$	σ^2

E Table of exponential families

Table E.6. exponential families: absolutely continuous distributions, part 4

Distribution	Log-normal	Rayleigh	Wald	Maxwell-Boltzmann
$f_X(x)$	$\frac{1}{x\sqrt{2\pi\sigma^2}}\exp\left(-\frac{|\log x-\mu|^2}{2\sigma^2}\right)$	$\frac{x}{\sigma^2}\exp\left(-\frac{x^2}{2\sigma^2}\right)$	$\sqrt{\frac{\lambda}{2\pi x^3}}\exp\left(-\frac{\lambda|x-\mu|^2}{2x\mu^2}\right)$	$\sqrt{\frac{2}{\pi}}\frac{x^2}{a^3}\exp\left(-\frac{x^2}{2a^2}\right)$
$x\in$	$[0,\infty)$	$[0,\infty)$	$[0,\infty)$	$[0,\infty)$
parameter	(μ,σ)	σ	(λ,μ)	a
$t(x)$	$(-\log^2 x,\log x)$	$-x^2$	$\left(-\frac{1}{x},-x\right)$	$-\frac{x^2}{2}$
θ	$\left(\frac{1}{2\sigma^2},\frac{\mu}{\sigma^2}-1\right)$	$\frac{1}{2\sigma^2}$	$\left(\frac{\lambda}{2},\frac{\lambda}{2\mu^2}\right)$	$\frac{1}{a^2}$
$F(\theta)$	$\frac{(\theta_2+1)^2}{4\theta_1}-\frac{1}{2}\log\theta_1$	$-\log(2\theta)$	$-\frac{\lambda}{\mu}-\frac{1}{2}\log\lambda$	$-\frac{3}{2}\log\theta$
$k(x)$	$-\log\sqrt{2\pi}$	$\log x$	$-\frac{3}{2}\log x$	$2\log x+\log\sqrt{\frac{2}{\pi}}$
$E(X)$	$\exp\left(\mu+\frac{\sigma^2}{2}\right)$	$\sigma\sqrt{\frac{\pi}{2}}$	μ	$2a\sqrt{\frac{2}{\pi}}$
$var(X)$	$(\exp(\sigma^2)-1)\exp(\sigma^2+2\mu)$	$(2-\frac{\pi}{2})\sigma^2$	$\frac{\mu^3}{\lambda}$	$a^2(3-\frac{8}{\pi})$

Table E.7. exponential families: absolutely continuous distributions, part 5

Distribution	Von Mises	U-power	Half normal	Inverse Chi square
$f_X(x)$	$\frac{\exp(\kappa\cos(x-\mu))}{2\pi I_0(\kappa)}$	$\frac{2k+1}{2\alpha}\left(\frac{x-\mu}{\alpha}\right)^{2k}$	$\sqrt{\frac{2}{\pi\sigma^2}}\exp\left(-\frac{x^2}{2\sigma^2}\right)$	$\frac{2^{-\frac{k}{2}}}{\Gamma(\frac{k}{2})}x^{-\frac{k}{2}-1}\exp(-\frac{1}{2x})$
$x\in$	$[\mu-\pi,\mu+\pi]$	$[\mu-\alpha,\mu+\alpha]$	$[0,\infty)$	$[0,\infty)$
parameter	(κ,μ)	k	σ	k
$t(x)$	$(\cos x,\sin x)$	$2\log(x-\mu)$	$-x^2$	$-\log x$
θ	$(\kappa\cos\mu,\kappa\sin\mu)$	k	$\frac{1}{2\sigma^2}$	$\frac{k}{2}$
$F(\theta)$	$\log I_0(\kappa)$	$2\theta\log\alpha-\log(\theta+\frac{1}{2})$	$-\log\theta-\frac{1}{2}\log 2$	$\frac{k}{2}\log 2+\log\Gamma(\frac{k}{2})$
$k(x)$	$-\log(2\pi)$	$-\log\alpha$	$\log\sqrt{\frac{2}{\pi}}$	$-\log x-\frac{1}{2x}$
$E(X)$	μ	μ	$\sigma\sqrt{\frac{2}{\pi}}$	$\frac{1}{k-2},k>2$
$var(X)$	$1-\frac{I_1(\kappa)}{I_0(\kappa)}$	$\frac{2k+1}{2k+3}\alpha^2$	$(1-\frac{2}{\pi})\sigma^2$	$\frac{2}{(k-4)(k-2)^2},k>4$

Solutions to selected problems

Chapter 1:

1.3 For part 3, use $p^{n+1} = p^n p$ and the fact that p is a Markov matrix. For part 4 note that the relation derived in part 3 shows that the sequence is monotone.

1.4 Note that $\{T_1 \geq n, X_0 = 0\}$ means that the chain stays stuck at 0; then use relation (1.1) and the Markov property.

1.5 You should recognize a geometric distribution.

1.7 For part 2, note that we always have $X_{n+1} > X_n$ as long as 5 is not reached, so that $1 \leq T \leq 4$. For each value k we can make a list of the possible trajectories starting at 1 to evaluate $P(T = k|X_0 = 1)$. For part 4, we may remark that the relation derived in part 3 shows that (formally, i.e., without worrying about convergence) for $j \leq N$, the quantity $E(N_j|X_0 = 1)$ is the $(1, j)$-entry of the matrix

$$R := I + p + \cdots + p^k + \cdots .$$

The last row of p has a 1 at the end, so when we compute p^k, we find the same block structure, with Q^k as the upper right block; this shows that $R(1, j) = S(1, j)$ if we define the matrix S by

$$S := I + Q + \cdots + Q^k + \cdots ;$$

which means $S = (I - Q)^{-1}$. Finally, the quantity of question 5 is the waiting time before reaching the absorbing state.

1.12 As usual, the extreme values $k = 0$ and $k = d$ need to be treated separately; if k is not an extreme value, consider the probability of drawing from the left (or right box), and see how k may change. This gives

$$p(k, k - 1) = \frac{k}{2d}, \; p(k, k) = \frac{1}{2}, \; p(k, k + 1) = \frac{d - k}{2d}.$$

© Springer International Publishing AG, part of Springer Nature 2018
J.-F. Collet, *Discrete Stochastic Processes and Applications*,
Universitext, https://doi.org/10.1007/978-3-319-74018-8

Solutions to selected problems

Chapter 2:

2.1 Note that since $y \neq 0$, we have $xy \neq 0$, and compute the product xAy.

2.3 Show that when one computes the successive powers of A, once a positive element appears, it remains forever: if $A^m(i,j) > 0$, then $A^r(i,j) > 0$ for all $r \geq m$.

2.5 Using the explicit formula for the inverse matrix (in which the determinant appears as a denominator), show that a 2×2 monotone matrix is an M matrix if and only if its determinant is positive.

2.6

1. The column vector $(1, \ldots, 1)^t$ should do the trick.
2. We know that for some vector $\bar{x} > 0$, the inequality $\bar{x} > A\bar{x}$ is satisfied; we can choose some $\lambda \in (0,1)$ such that $\lambda \bar{x} > A\bar{x}$ (note how the fact that the inequality is strict plays an essential role here). Then show that for every n we have $\lambda^n \bar{x} > A^n \bar{x}$, and deduce that $A^n \bar{x} \to 0$. Finally, use the fact that $\bar{x} > 0$.
3. Let $\bar{x} > 0$ be as above; pick x such that $(I - A)x \geq 0$. We need to show that $x \geq 0$. Argue by contradiction and consider the set

$$A := \{\alpha \in \mathbb{R} : \quad x + \alpha \bar{x} \geq 0\}.$$

   Show that A is an interval of the form $[\alpha_0, \infty)$ for some $\alpha_0 > 0$, and get a contradiction by showing that $x + \alpha_0 \bar{x} > 0$.

2.7 Take a look at the justification of Theorem C.4 based on the Schur form of A.

Chapter 3:

3.2 If we use the method of test functions and compute $E(\phi(T_n))$, then Fubini's formula gives us a sequence of $n - 1$ one-dimensional integrals, which may be computed sequentially.

3.3 Each jump of X being exactly of magnitude 1, you may use the equality of events

$$\{X_t \geq n\} = \{T_n \leq t\}$$

to compute $P(T_n \leq t)$ and then differentiate.

3.4 Use the equality of events $\{E_t < s\} = \{T_{X_t} > t - s\}$ and treat the cases $s \geq t$, $s < t$ separately to compute $P(E_t \leq s)$.

3.5 We want to check Definition 3.8; after having checked stationarity and independence of the increments, use total probability (conditioning on the result of α) to determine the distribution of Z_t.

3.6 Using the definition of conditional probability, independence, and stationarity of the increments, show that for $k \leq n$,

$$P(X_s = k | X_t = n) = P(X_{t-s} = n - k)\frac{P(X_s = k)}{P(X_t = n)},$$

and then compute the right-hand side.

3.7 Use total probability (conditioning by the value of X_t), and to compute the sum, note that conditional on $X_t = n$, the variable X_t^1 has the binomial distribution $\mathcal{B}(n, p)$.

3.8 To find the correlation of X_t and X_s you may determine $E[(X_t - X_s)^2]$ and then deduce the value of $E[X_t X_s]$.

3.9 The expectation is computed using the law of total probability (conditioning by $N_t = n$); then the expectation of Z_t is obtained by taking the first derivative $\Phi'_{Z_t}(0)$.

3.10 The number of times X_t a fixed ball has changed boxes between 0 and t is $\mathcal{P}(\lambda t)$, and $p(t)$ is the probability that X_t is even. If all balls are initially in the left box, then $N(t) \sim \mathcal{B}(d, p(t))$; for the large-time asymptotics, note that $p(t) \to \frac{1}{2}$. For the general case in which the initial number of balls in the left box is j, we are going to distinguish among the balls present in the left box at time t those coming from the left from those coming from the right: $N(t) = N^l(t) + N^r(t)$, where $N^l(t) \sim \mathcal{B}(j, p(t))$ and $N^r(t) \sim \mathcal{B}(d - j, 1 - p(t))$ are independent. In the limit $t \to \infty$, the limit is given by

$$\lim_{t \to \infty} p_k(t) = \frac{1}{2^d} \sum_{l=0}^{k} \binom{j}{l}\binom{d-j}{k-l} = \frac{1}{2^d}\binom{d}{k},$$

the second equality being a consequence of (B.3); so again the limiting distribution is $\mathcal{B}(d, \frac{1}{2})$.

3.11 For X_1^t use the equality of events

$$\{S_1^t > s, X_t = n\} = \{X_{t+s} = n, X_t = n\}.$$

For the third question, the strategy is to formulate everything in terms of the interarrival times S_i; note that $\{X_t = n\} = \{T_n \leq t < T_n + S_{n+1}\}$, and

$$\{S_1^t > s_1, X_t = n\} = \{t + s_1 < T_n + S_{n+1}, X_t = n\}$$

(drawing a picture might help). Therefore, by independence of interarrival times, we have

$$P(S_1^t > s_1, \ldots, S_k^t > s_k, X_t = n\})$$
$$= P(T_n \leq t, t + s_1 < T_n + S_{n+1})P(S_{n+2} > s_2) \cdots P(S_{n+k} > s_k). \quad (1)$$

The first term on the right-hand side may be computed using the joint density of T_n and S_{n+1} (just as we did in the proof of Theorem 3.16), and the conclusion follows by summing over $n \in \mathbb{N}$.

Solutions to selected problems

Chapter 4:

4.2 For $s < t$, use transitions over the time intervals $[s+h, t+h]$ and $[s, t]$ to write

$$p_{ij}(t+h) - p_{ij}(t) = \sum_k [p_{ik}(s+h) - p_{ik}(s)]p_{kj}(t-s),$$

and then bound this quantity from above.

4.3 Note that over a given time interval $[0, t]$, the value of Y changes if and only if X_t is odd; use this to derive the transition matrix (on the space $\{-1, 1\}$)

$$p_t = \frac{1}{2} \begin{pmatrix} 1 + \exp(-2\lambda t) & 1 - \exp(-2\lambda t) \\ 1 - \exp(-2\lambda t) & 1 + \exp(-2\lambda t) \end{pmatrix}.$$

4.4 Note that every 2×2 Markov matrix p_t is of the form chosen here; in the forward as in the backward case, there are only two independent differential equations; then by Corollary C.21, both problems are equivalent if and only if the matrices A and $p(0)$ commute, and this is exactly equivalent to the required condition. For a direct proof note that the first system implies (since v' is a multiple of u') that $v - \frac{\mu}{\lambda}u$ is a constant; in other words, we have

$$v = v_0 - \frac{\mu}{\lambda}u_0 + \frac{\mu}{\lambda}u,$$

and (after some basic manipulations) the first system assumes the following form:

$$\begin{cases} u'(t) = -(\lambda + \mu)u - \lambda v_0 + \mu u_0 + \lambda, \\ v'(t) = -(\lambda + \mu)v + \lambda v_0 - \mu u_0 + \mu, \end{cases}$$

which may then be compared to the second system.

4.5 For the last question use the equivalent $\ln(1+h) \sim h$ for h near 0.

4.6 The key point here is that although we have no uniformity condition (compare to Theorem A.8, for instance), the tail of the series may be expressed in terms of the whole sum and the truncated sum. To be more specific, let $|v|_\infty := \sup_n v_n$; writing

$$\sum_{n=0}^N u_n(t)v_n \leq \sum_{n=0}^\infty u_n(t)v_n \leq \sum_{n=0}^N u_n(t)v_n + |v|_\infty \left(\sum_{n=0}^\infty u_n(t) - \sum_{n=0}^N u_n(t) \right),$$

you may then let $t \to 0^+$ and conclude as in the proof of Theorem 4.18.

Chapter 5:

5.2 For the Taylor series of f, show by induction on k that for $k \geq 2$,

$$f^{(k)}(x) = -2^k [3 * 5 * \cdots * (2k-3)](1 - 4x)^{-\frac{2k-1}{2}}.$$

The formula for the quotient of factorials may be proved by induction, or (better) directly by noting the cancellation of all even factors in the numerator. Finally, just compute $\frac{1-f(x)}{2x}$; for completeness, you may even check continuity of the result at 0.

5.3 Begin by showing that if x, y are two integers with $x < y$, then

$$x!y! \geq (x+1)!(y-1)!;$$

then for the general case, use a "perturbation" argument: if some of the n_i are far from $[\frac{n}{d}]$, you may decrease the large ones and increase the small ones to produce a lesser value of the product. For the case $n = 2$, the result (about the uniform binomial distribution) says that the binomial coefficient $\binom{n}{k}$ is maximal when k is near $[\frac{n}{2}]$; note that you may also prove this fact by studying the monotonicity of the sequence $u_k := \binom{n}{k}$ for k between 0 and n. For the multinomial distribution it says that the most probable distribution is that in which all boxes contain (as close as possible to) the same number of elements.

5.4 This is simply the probability p that a path drawn uniformly on the set of $2n$ loops is a simple loop; after simplification one obtains $p = \frac{1}{2n-1}$.

5.6 To recover the Poisson process, note that if $\mu = 0$, the ordinary differential equation is immediately integrated to give

$$\phi(z, t) = e^{\lambda(z-1)t} \phi(z, 0).$$

To obtain $p_t(j, \cdot)$ we take the initial data $u_i(0) = \delta_{ij}$; thus $\phi(z, 0) = z^j$. We may then expand $\phi(z, t)$ in powers of z:

$$\phi(z, t) = z^j e^{\lambda(z-1)t} = e^{-\lambda t} \sum_{l=0}^{\infty} \frac{(\lambda t)^l}{l!} z^{l+j} = e^{-\lambda t} \sum_{k=j}^{\infty} \frac{(\lambda t)^{k-j}}{(k-j)!} z^k.$$

By identification of the general term in ϕ this gives the expression for $u_k(t) = p_t(j, k)$.

Chapter 6:

6.1 Parametrize the segment $[x, y]$ and apply the one-variable formula.

6.3 Express $Ap \cdot p$ as a double sum and symmetrize it. In other words, write

Solutions to selected problems

$$A = \sum_i \sum_j w_{ij} = \sum_i \sum_j w_{ji} = \frac{1}{2} \sum_i \sum_j [w_{ij} + w_{ji}].$$

6.4 Showing that the lower bound obtained in Problem 6.3 is sharper (which means larger) amounts to showing that for all P, Q positive, one has

$$P \log \frac{P}{Q} \geq P - Q.$$

6.6 Use the same symmetrizing trick as in Problem 6.3.

6.8 The relation between $\det A$ and $\det C$ comes from multilinearity of the determinant (spot common factors in each row and each column of A); to get the relation between D_n and D_{n-1}, perform a row manipulation on D_n. This will give an expression of $D_n + D_{n-1}$ as a determinant; perform a row manipulation on this determinant to get the result.

6.9 The symmetrizing trick of Problem 6.3, yet again!

6.13 To obtain (6.23), apply (6.22) to $u - \int u\phi$ and $v - \int v\phi$; then put $\phi := \frac{b}{B}$ in (6.23) to obtain (6.24).

Chapter 7:

7.3 For part 2, use the representation of P as a convex combination of permutation matrices and the fact that H is concave; for part 3, take p uniform; for part 4, use inequality (6.19) with exponents $P(x, y)$.

7.7 For the first question, simply apply Jensen's inequality to the given map, writing $\frac{q_i}{p_i} = \exp(-\ln \frac{p_i}{q_i})$; to get $D_{KL}(p, q)$ large, take, for instance, two Bernoulli variables of parameters p and q, with p fixed and $q \to 0$; finally, the absolute bound for D_{JS} is an immediate consequence of $D_J(p, q) \geq 0$.

7.8 Express $H(X_1, \ldots, X_n)$ by leaving X_n aside and conditioning; then use Cesaro's lemma.

7.10 The first relation is immediate from the Fokker–Planck equation; for the second one, use the following summation by parts formula:

$$\sum_{k=1}^{\infty} a_k(b_{k+1} - b_k) = -a_1 b_1 - \sum_{k=1}^{\infty} b_{k+1}(a_{k+1} - a_k).$$

Chapter 8:

8.6 For part 3, note that z^x increases with x if and only if $z > 1$; thus we are interested in having $\frac{r}{1-r} < 1$, which means $r < \frac{1}{2}$.

References

1. Emil Artin. 2015. *The Gamma function*. Mineola: Courier Dover Publications.
2. Robert B Ash. 1990. *Information theory*. New York: Dover publications.
3. Monica Bianchini, Marco Gori, and Franco Scarselli. 2005. Inside PageRank. *ACM Transactions on Internet Technology (TOIT)* 5 (1): 92–128.
4. Leo Breiman. 1992. *Probability, volume 7 of classics in applied mathematics*. Philadelphia: SIAM.
5. Pierre Brémaud. 2013. *Markov chains: Gibbs fields, Monte Carlo simulation, and queues*, vol. 31. Berlin: Springer Science & Business Media.
6. Richard A Brualdi. 1982. Matrices eigenvalues, and directed graphs. *Linear and Multilinear Algebra* 11 (2): 143–165.
7. Thomas M Cover, and Joy A Thomas. 2012. *Elements of information theory*. New York: Wiley.
8. GE Crooks. 2008. Inequalities between the Jensen–Shannon and Jeffreys divergences, tech. Technical report, Note 004, 2008. http://threeplusone.com/pubs/technote/CrooksTechNote004.pdf.
9. N Dmitriev, and E Dynkin. 1945. On the characteristic numbers of a stochastic matrix cr acad. *URSS (NS)* 49: 159–162.
10. Nikolai Aleksandrovich Dmitriev, and E Dynkin. 1946. On characteristic roots of stochastic matrices. *Izvestiya Rossiiskoi Akademii Nauk. Seriya Matematicheskaya* 10 (2): 167–184.
11. Richard Durrett. 2012. *Essentials of stochastic processes*. Berlin: Springer Science & Business Media.
12. Georg Ferdinand Frobenius. 1912. *Über Matrizen aus nicht negativen Elementen*. Königliche Akademie der Wissenschaften.
13. Jean-Baptiste Hiriart-Urruty, and Claude Lemaréchal. 2012. *Fundamentals of convex analysis*. Berlin: Springer Science & Business Media.
14. Morris W Hirsch, Stephen Smale, and Robert L Devaney. 2012. *Differential equations, dynamical systems, and an introduction to chaos*. Academic press.

© Springer International Publishing AG, part of Springer Nature 2018
J.-F. Collet, *Discrete Stochastic Processes and Applications*,
Universitext, https://doi.org/10.1007/978-3-319-74018-8

References

15. Roger A Horn, and Charles R Johnson. 2012. *Matrix analysis.* Cambridge: Cambridge university press.
16. Gareth A Jones, and J Mary Jones. 2012. *Information and coding theory.* Berlin: Springer Science & Business Media.
17. Amy N Langville, and Carl D Meyer. 2004. Deeper inside PageRank. *Internet Mathematics* 1 (3): 335–380.
18. Amy N Langville, and Carl D Meyer. 2011. *Google's PageRank and beyond: The science of search engine rankings.* Princeton: Princeton University Press.
19. Gregory F Lawler, and Vlada Limic. 2010. *Random walk: a modern introduction,* vol. 123. Cambridge: Cambridge University Press.
20. Erich L Lehmann, and Joseph P Romano. 2006. *Testing statistical hypotheses.* Springer: Springer Science & Business Media.
21. Erich Leo Lehmann, and George Casella. 1998. *Theory of point estimation,* vol. 31. Berlin: Springer Science & Business Media.
22. David JC MacKay. 2003. *Information theory, inference and learning algorithms.* Cambridge: Cambridge university press.
23. Marvin Marcus, and Henryk Minc. 1992. *A survey of matrix theory and matrix inequalities,* vol. 14. Courier Corporation.
24. Robert McEliece. 2002. *The theory of information and coding.* Cambridge: Cambridge University Press.
25. Carl D Meyer. 2000. *Matrix analysis and applied linear algebra,* vol. 2. Philadelphia: Siam.
26. James R Norris. 1998. *Markov chains.* Cambridge: Cambridge university press.
27. Fritz Oberhettinger, and Larry Badii. 2012. *Tables of Laplace transforms.* Berlin: Springer Science & Business Media.
28. Lawrence Page, Sergey Brin, Rajeev Motwani, and Terry Winograd. 1999. The PageRank citation ranking: Bringing order to the web. Technical report, Stanford InfoLab.
29. Oskar Perron. 1907. Zur theorie der matrices. *Mathematische Annalen* 64 (2): 248–263.
30. Edward M Reingold. 2000. Mathematical entertainments: Cliques, the Cauchy inequality, and information theory (department). *The Mathematical intelligencer* 22 (4): 14–15.
31. Ralph Tyrell Rockafellar. 2015. *Convex analysis.* Princeton: Princeton university press.
32. Walter Rudin. 1991. *Functional analysis,* International series in pure and applied mathematics. New York: McGraw-Hill Inc.
33. Joel L Schiff. 2013. *The Laplace Transform,* Undergraduate Texts in Mathematics. New York: Springer.
34. Richard P Stanley. 1999. *Enumerative combinatorics,* vol. 2, vol. 62 of cambridge studies in advanced mathematics. Cambridge: Cambridge University Press.

35. Richard S Varga. 2001. Gerschgorin disks, Brauer ovals of Cassini (a vindication), and Brualdi sets. *Information-Yamaguchi* 4 (2): 171–178.
36. Tomasz Zastawniak, and Zdzislaw Brzezniak. 2003. *Basic stochastic processes*. Berlin: Springer.

Index

Index

Index

Printed in the United States
By Bookmasters